EXTERRANEAN

MEANING SYSTEMS

EXTERRANEAN

Extraction in the Humanist Anthropocene

PHILLIP JOHN USHER

Fordham University Press: New York 2019

Visit us online at www.fordhampress.com.

Library of Congress Control Number: 2018961515

Printed in the United States of America

21 20 19   5 4 3 2 1

First edition

for you, mum

CONTENTS

FIGURES

What the Earth has hidden and kept underground . . . destroys us.
—Pliny, *Natural History*

INCIPIT

From Sub- to Exterranean

> . . . the voice-less things once placed as a décor surrounding the usual spectacles, all those things that never interested anyone, from now on thrust themselves brutally and without warning into our schemes and maneuvers.
>
> —Michel Serres, *Le contrat naturel*

> Humans in the sixteenth century did not observe things with a cold eye, with a detached gaze. They felt wholly bound up with them. They are voices with which all inquiry and all exchange is a veritable dialogue.
>
> —Jean Céard, *La nature et les prodiges*

This is a book about the extraction of *stuff* from the Earth, a process in which matter goes from being *sub-* to *ex*terranean.[1] Its driving conviction is that if we have stumbled into the Anthropocene, this age in which humans are the major geological re-shapers of Planet Earth, it is not only because we *emit* but first and foremost because we *extract*. Much of the CO_2 that now fills our atmosphere, currently at a concentration of about 400 parts per million, was released by the combustion of *stuff* (coal, oil, gas, etc.) that used to be underground and materially connected to the Earth.[2] While we *know* this, we clearly do not always *feel* or *remember* it. The document known as the Paris Agreement, produced within the UN Framework Convention on Climate Change (COP 21) in December 2015, is thirty-one pages long. It uses the word *emission* a total of ninety-eight times. Words such as *extraction* and *mining* appear not at all, as if what we burn had no origin. An exercise in ecological mindfulness, this book's aim is to make perceptible the material and immaterial entanglements of the Earth from which we extract, the human and nonhuman agents of extraction, and the extracted matter with which we live daily. In crafting the concept of the exterranean, my hope—and here I follow the recent work of Bruno Latour—is to produce a shift away from the idea that to think at the scale of the planet in the Anthropocene is to keep in our mind the figure of a globe. We can no longer make do with looking back at Earth *as if from nowhere* or *as if from outer space*, the kind of global thought associated with Lucian's *Icaromennipus*, the inventions of early modern cartographers, and most famously with the *Earthrise* photo.[3] In place of the globe's totalized Earth, this book presents Earth relationally: We live not just *on* the Earth but *with* it.[4]

If to think of the Earth as a globe means to take oneself for God, as Latour argues it is, then we should instead seek out a sense of planetary scale that is obtained via "the ability to establish more or less numerous and most importantly reciprocal relations"; that is, we should follow "feedback loops" and "avoid totality."[5] This is why in his most recent work Latour has turned to explore Earth's critical zone, that "spot on the envelope of the biosphere" that "extends *vertically* from the top of the lower atmosphere down to the so-called sterile rocks and *horizontally* wherever it is possible to obtain reliable data on the various fluxes of ingredients [i.e., all kinds of elements, whether chemical, physical, political, etc.] through the chosen site."[6] Such a turn was central to Latour's 2016 exhibition *Reset Modernity!* at the Zentrum für Kunst und Medien Karlsruhe. In a room titled "From Land to Disputed Territories," a sign reads: "Instead of looking at the 'blue planet,' what about digging through critical zones, examining the thin planetary membrane that contains all forms of living beings?"[7] The room responds by juxtaposing inter alia Pierre Huyghe's *Nymphéas transplant (14–18)*, an aquarium inspired by—or better, recomposed after—Monet's famous *Nymphéas* painting and maps detailing "the extensive footprint of oil extraction in Texas, Canada, the Middle East, Nigeria, and the Arctic,"[8] in order to direct attention away from the *globe* on which we live but which we can never truly see and toward the critical zones on which life depends and with which we clearly do interact each and every second. This concern—to eschew globes and the Global (to which Latour gives a capital G) and to avoid the equally troublesome Local (also capitalized)—is also central to, and further developed in, Latour's latest book *Où atterrir? (Where to Land?)*, which came into print as I finalized *Exterranean*. There, Latour argues against the Global, which "apprehends all things *from afar*, as if they were *exterior* to the social world and quite *indifferent* to human concerns," and in favor of the terrestrial (*le terrestre*), which "apprehends the same assemblages [*agencements*] as if seen *up close*, *interior* to collectives and *sensitive* [*sensibles*] to the action of humans to whom they keenly react."[9] *Exterranean* is in many ways an enterprise parallel to Latour's turn to critical zones and to the *terrestrial* and might be seen as anticipating or responding to *Où atterrir?*'s call for new descriptive cartographies of the terrestrial, although its scope is more limited—in what follows, I am only interested in the material and immaterial, visible and invisible, knottings of the Earth and the "stuff" that gets disconnected from it.[10]

Before we go any further, it is worth asking: If there is a need to rephenomenalize extraction, then why also a new word? How does it help to talk about the exterranean? My intention is that the new term respond to a need to grasp at, in just one word, all of the following: the land/ground/place where extraction occurs (*terra*), the planet to which this land/ground/place belongs (*Terra, tellus*),[11] the action of moving matter away from the land/ground/place/planet (*ex*), and that matter itself (which, removed, becomes—and always will be, can never not be—*exterranean*). In opposition to this word, the terms "mining" and "extraction" create a cut and make that cut invisible. To mine (< OFr. *miner* = to dig under land, a rock, so as to make it collapse; cf. to undermine) captures the moment of collapse—*mines* are, etymologically, places for disintegration.[12] To extract (< Lat. *extrahere* < *ex-* [out] + *trahere* [to draw]) emphasizes the action of removal. Both terms (*to mine, to extract*) perform a rupture: The ground/hillside/planet becomes raw material/product, such that what is one moment merely a part of the Rhondda Valley is transformed, materially and ontologically, into coal. The words *to mine* and *to extract* make us forget where that coal comes from. To talk of the exterranean, on the contrary, allows us to think-feel material continuities and to take into hermeneutic custody all of the human and nonhuman agents and materials of the process. This is, theoretically, an important move, for it privileges neither the miner, nor the pickaxe, nor the valley, nor the firedamp, nor the gold, nor the ring wearer. The term is capacious, and for that reason—and it is theoretically productive this way—what it designates is harder to access. It is possible to point to a miner's pickaxe and conclude: "At this precise moment, a miner is extracting marble from this quarry." We can, still easily, call this exterranean activity, for matter is being removed from (*ex*) the ground (*terra*). It is not possible, however, when we nominalize the adjective, to point to, to localize, the exterranean—just as we can watch a *film* but not see *the filmic*. Accessing the exterranean thus requires multiple entry points and perspectives, renewed attempts to perceive connections that we normally overlook.

To prepare fully for exploring the consequences of this proposed conceptual leap, let us first refashion the *OED*'s entry for *subterranean*, substituting for *sub* the preposition *ex* + ablative (out from, away) and allowing *terra* to be infiltrated with *Terra*, as if the term were now to enter the dictionary:

1. a. Of a physical phenomenon, force, movement, etc.: operating or performed moving away from the earth; occurring out of or from the ground/the planet.
   b. Of an inanimate object: existing, lying, or situated apart from the ground/planet; formed or constructed by coming from, by being disconnected from, the ground/the planet, either naturally or by human activity.
   c. Of a person, animal, etc.: constituted from, or living or working in a movement away from, the ground/planet.
   d. *Bot.* Of a plant, part of a plant, fungus, etc.: growing of/away from the ground.
2. Existing, belonging to, or characteristic of a distancing, or distanced, relationship to hell or the underworld; partaking of, but separate from, the infernal.
3. *fig.* Existing or working in a manner that is at a certain remove from the clandestine, not quite hidden.

To summarize: Exterranean, then, are those physical phenomena that perform an action in a direction that leads away from the ground/planet (e.g., mining, extraction, but also the accountancy of a mining company, whatever other activities contribute to the possibility of mining, etc.); inanimate (but perhaps vibrant) objects that were, but no longer are, part of the ground/planet (e.g., coal, gold, iron, etc.); animate beings that are constituted from, or who live, work, or move, in a relationship of growing distance from the ground/planet; plants that grow from the ground/earth. By extension, exterranean is what exists, belongs to, or is characteristic of moving away from the infernal underworld. And figuratively—if we want this opportunity—exterranean is what exists or works in a nonhidden manner. A book titled *Exterranean* will thus logically strive to account for a lot of realities and phenomena, and it must logically do so from multiple, and not necessarily a priori reconcilable, perspectives.[13]

THE HUMANIST ANTHROPOCENE

In what follows, I approach and develop the notion of the exterranean from what might appear at first to be an oblique perspective, by engineering collisions between the reality and theory of our present moment and texts and images from early modern Europe, a collision laboratory that I refer to as the Humanist Anthropocene. Because the whole book happens in this intellectual sandbox, a brief aside here is essential, before returning to the exterranean in particular. I shall thus be putting, because of their differences and *différends*, the *anthropos* of the Anthropocene in dialogue with the *homo* of humanism, a dialogue that aims to respond to Dipesh Chakrabarty's central insight: "However anthropogenic" the situation of the Anthropocene may be, "there is no corresponding 'humanity' that in its oneness can act as a political agent."[14] I expect that both scholars of early modern Europe and theorists of the Anthropocene might, on first blush, diagnose here a bad case of anachronism. Surely, both might say, the Anthropocene calls for posthumanism. To appreciate the work that the proposed encounter might allow, we must begin not by defining humanism and the Anthropocene but by untranslating—as I have done in more detail elsewhere—the terms *homo* and *anthropos* from which they are hewn.[15] Both *appear* to refer to something like "man" or "human," but they actually apprehend quite different aspects of the human and populate our academic lexicon in vastly different ways. On the one hand, there is the *anthropos* of anthropology—s/he is an object of study, traditionally marked as foreign/other, and defined in opposition to the traditionally Western, notebook-wielding academic. We find the same reified *anthropos* in anthropometry, the science of measuring the human body—s/he is not the hand that measures but the ear caught between the calipers (Figure 1).[16] The extent to which this observer-measurer might, in either case, share humanity with the studied *anthropos* is uncertain, put in parentheses.

The *anthropos* of anthropology and anthropometry, the observed, measured, out-there-and-not-quite-me human, reappears in and constitutes the Anthropocene—from the very moment Eugene Stoermer, a professor of biology, and Paul Crutzen, the Nobel Prize–winning atmospheric chemist, coin and popularize the term, which is precisely why the Anthropocene raises, as many thinkers have noted, such complex questions about agency, about who *we* are, about how *we* relate to what *we* have done. We can observe this distancing of the human in Stoermer's and Crutzen's very first (and very short) article about the Anthropocene, published in *Global Change Newsletter* in May 2000. Take the following extracts from that article, which I list here in the order they appear:

The expansion of *mankind*, both in numbers and per capita exploitation of Earth's resources, has been astounding.

In a few generations, *mankind* is exhausting the fossil fuels that were generated over several hundred million years.

More than half of all accessible fresh water is used by *mankind*.

Human activity has increased the species extinction rate by [a] thousand to ten thousand fold in the tropical rain forest.

Mankind releases many toxic substances in the environment.

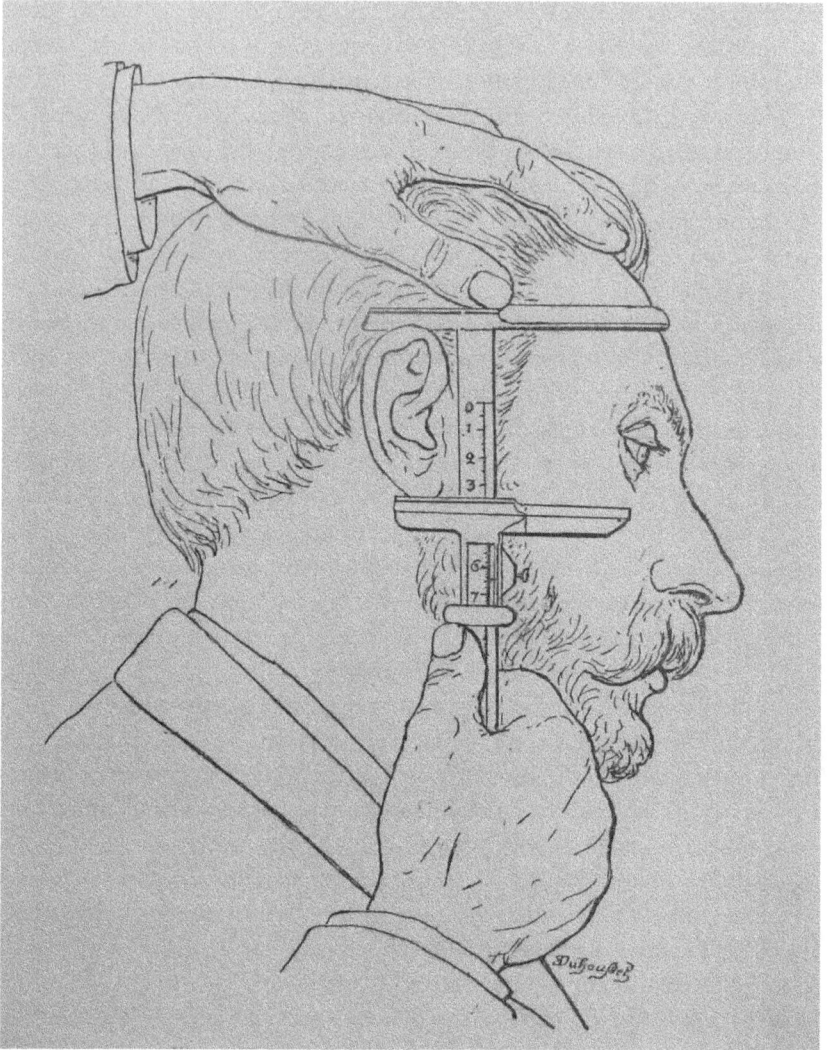

FIGURE I. "Measuring the Ear of the Anthropos," from Alphonse Bertillon, *Identifi-cation anthropométrique* (1893). Division of Rare and Manuscript Collections, Cornell University Library.

Coastal wetlands are also affected by *humans*, having resulted in the loss of 50% of the world's mangroves.

Mechanized *human* predation ("fisheries") removes more than 25% of the primary production of the oceans in upwelling regions and 35% in the temperate continental shelf regions.

Anthropogenic effects are also well illustrated by the history of biotic communities that leave remains in lake sediments.

It seems to us more than appropriate to emphasize the central role of *mankind*
in geology and ecology by proposing to use the term *"anthropocene"* for
the current geological epoch.[17]

Over the course of their short article, the two scientists place the greatest
emphasis on *mankind* (i.e., *Menschheit*), turning quickly to Greek to speak of
*anthropo*genic effects and finally of the Anthropocene. Like practitioners of
anthropometry, they seek to understand and to measure—from a distance
of scientific analysis—an alienated and strange human. If we imagine graphs
that chart a variety of anthropogenic changes wrought and written on Planet
Earth—atmospheric CO_2 concentration, great floods, the fall in global biodiversity, the loss of tropical rainforest and woodland, ozone depletions, as well
as correlated measures of human activity: population, GDP, water use, number
of motor vehicles, McDonald's restaurants, international tourism—we see a
callipered, averaged, abstract human that is not us. Our hands turning the
keys of our car ignitions are there, somewhere, within those averages, but
completely impossible to locate. As I am far from the first to note, the *anthropos* of the Anthropocene is thus everyone and no one, all of us and none of
us, a guilty party to whom it is difficult to assign agency—and first and foremost, a human observed and aggregated by an agency whose humanness is ill-
defined.

On the other hand, there is the *homo* of humanism, a word that, having been
around a lot longer than "Anthropocene," possesses an even greater ability to
befuddle. Let's be clear from the outset about two points. First, the humanism of the Humanist Anthropocene is *early modern* humanism. Second, this
homo is neither our hero nor our savior, just a second, different, human. Before
directly addressing the second point, and given that a certain understanding
of the word "humanism" all too often these days becomes posthumanism's
mystifying *faire-valoir*, a problem of evident consequence for this book's own
suturing of the early modern and the contemporary, we must attend to the
first point and unravel the term.[18] As Kenneth Gouwens has shown recently in
an urgent article titled "What Posthumanism Isn't," Cary Wolfe in his *What Is
Posthumanism?* pays "negligible attention" to early modern humanism—there
is no mention of Petrarch, Valla, Erasmus, or others in Wolfe's book. No study
can be exhaustive, of course. Problematic, however, as Gouwens demonstrates
in no uncertain terms, is that despite this lack of coverage, "the humanism
that Wolfe sets up as a foil resembles only marginally how leading scholars of
intellectual and cultural history have used the term with respect to Europe

in the period extending roughly from 1250 to 1600."[19] Gouwens's admonition is surely justified: Wolfe opens his book by stating that "[most] definitions of humanism look something like the following one from *Wikipedia*: 'Humanism is a broad category of ethical philosophies that affirm the dignity and worth of all people,'" to which he opposes posthumanism as a counterconcept capable of "decentering of the human."[20] As Gouwens phrases it, Wikipedia, while "as good a place as any to eavesdrop on the *Zeitgeist*," here offers up a definition that is "wildly misleading."[21] Rosi Braidotti similarly, in her *The Posthuman*, also emphasizes the idea of a "de-centering of Man" as constitutive of post-humanism and—as if to stage this decentering as a simultaneous rejection of specifically early modern humanism—she opposes Da Vinci's *Vitruvian Man* (c. 1490) to various new artworks that rework it, such as Maggie Stiefvater's *Vitruvian Cat*, to encapsulate the idea that early modern humanism has been superceded by posthumanism.[22]

Such stories—even as they proclaim to be theoretical rather than strictly historical—are disingenuous. For sure, some early modern humanists seem to celebrate human excellence, anthropocentrism, or human exceptionalism— but such celebrations are not universal and certainly *never* constitutive of humanism; moreover, they frequently result from misreadings of key texts, most notably of Pico della Mirandola's so-called *De hominis dignitate* (*On the Dignity of Man*).[23] If one needs a memorable reminder that early modern humanists could—just like posthumanist thinkers—decenter the human, one need only think of Montaigne's quip in the "Apologie de Raimond Sebond" ("Apology of Raymond Sebond"), part of course of a much longer and more complex argu-ment, that "[la] plus calamiteuse et fraile de toutes les creatures, c'est l'homme, et quant et quant la plus orgueilleuse" ("[the] most wretched and frail of all creatures is man, and at the same time the most puffed up with pride").[24] The point is not just that Montaigne and many other early modern humanists were already posthumanists in many regards but rather that to oppose posthuman-ism to early modern humanism is simply erroneous and confusing. It is impor-tant to remember that the popular—let's call it Wikipedian—understanding of humanism as some cocktail of secular anthropocentrism and generalized empathy for humanity has a long and complicated history. Vito R. Giustiano plots that history back to eighteenth-century France, when a 1765 journal article reasoned that the "general love for humanity" might be called human-ism, "for it is high time that a word be created for such a beautiful and neces-sary idea."[25] After the various senses given to the word by Marx, Feuerbach, Heidegger, and others, the 1933 *Humanist Manifesto*, signed by thirty-four intel-lectuals including John Dewey, would secure for good the idea of humanism as a secular celebration of human potential, "replacing traditional religious beliefs by stalwart confidence in our capability to achieve moral perfection and happiness."[26]

Early modern humanism, then, is something different from the 1765 and 1933 versions and certainly not a synonym for anthropocentrism. For some first definitions about it and the *homo* who stands in opposition to the *anthropos* it can be asserted that *this* human is not an object of study (as is the *anthropos*) but rather the subject who does the studying. This changes everything. In classical Latin, *humanus* had two complementary meanings: As the adjective of *homo*, it indeed could just refer to someone as a member of a species, that is, German *Menschheit*, but it could also refer to those (acquired, not innate) qualities, especially learnedness, that marked a human being as fully embracing the possibilities of humanity via study, that is, German *Humanität*.[27] Thus Cicero spoke, in the *Pro archia*, of the *studia humanitatis* (i.e., the ancestor of the humanities) to refer to "all of those related arts which one studies for the first time in . . . youth [in order to] attain *humanitas*."[28] We see here that the *anthropos* and the *homo* part ways, as Latin *humanitas* comes to betoken not an eternal and shared humanness but something additional and that separates the human not from the divine but from the uncivilized and the unlearned. There is the bare *homo*, and then there is the *humanus homo*—this is why it was not an uncommon compliment to refer to an esteemed author as *humanissime vir*, that is, most worthy (i.e., most human) man.[29] The productive tension between the *homo* and the *anthropos* crystallizes in that famous line from Terence's *Self-Tormentor*: "I am a human [*homo sum*] and, I think, nothing pertaining to humans is foreign to me [*humani nil a me alienum puto*]," a line said by Augustine to have been met by great applause on its first being spoken and that Montaigne inscribed on one of the beams in his tower library.[30] Terence's formula asserts an earnest and direct connection between the thinking human (*homo sum/puto*) and the general category of the human (*humani*), between which nothing (*nihil*) can come. The line says nothing less than the exact opposite of the Anthropocene, which announces rather: *I look on at what humans have done, and it and they remain wholly alien to me.* One figure of the human beckons alienation; the other does not. This is the *homo*'s potential within the Humanist Anthropocene, then: to provide a subject to the *anthropos*'s object.[31]

The *subjectness* of the *homo* is constitutive of early modern humanism. It is indeed symptomatic of this that before the word *humanism* came—in the 1840s, via the writings of Karl Hagen and Georg Voigt, to connect explicitly the term humanism and the early modern period[32]—it was first the term *humanista* that entered circulation in fifteenth-century Italy to name a teacher of *humanae litterae*, that is, texts of classical antiquity.[33] Before a movement or a corpus, there were thinking subjects. What interests me in these particular thinking-reading-translating-writing subjects is—and the weighing up of this hypothesis is precisely the subject of the chapters that follow—that they are not modern in the sense that Bruno Latour gives to that term. The founding intuition of the Humanist Anthropocene is that early modern humanists read and wrote with a sense of what Latour calls "analytic continuity" (*la continuité des analyses*), which the project of modernity largely dismantled to the point that "the representation of nonhumans" came to belong "to science" while science came to be no longer "allowed to have any relation to the nonhumans produced and mobilized by science and technology."[34] The beforeness of early modern humanists—they read and wrote *before* modern science, *before* modern geology, *before* Descartes and Bacon, etc.—is what makes them so urgently *not modern* and thus so potentially helpful in thinking the Anthropocene. I am not the first to make such an assertion, of course: Louisa Mackenzie, for one, also building on Latour, writes that "if we have never been modern, perhaps we have always been early modern."[35] As Mackenzie also argues, early modern humanists—in which category, by definition and following Anthony Grafton, are also included scientists—"were Latourian thinkers" in that their understanding of *physis*—and this is surely the meaning of the epigraph from Jean Céard—was "always already philosophical and theological"; their *natura* (whether *natura naturans* or *natura naturata*) took for granted that—and here Mackenzie quotes Jeffrey Theis—"culture is inseparable from the materiality of nature."[36] The Humanist Anthropocene thus hopes to be an experimental concept-ground in which this not-modern subject *homo* dialogues with the object *anthropos* of our times.

HUMANIST MINING

If the present book asserts that the humanists of early modern Europe are, as "Latourian thinkers" (in Mackenzie's formulation quoted above), invaluable for articulating the notion of the exterranean, it is also because some of them were the authors of the very first books to deal explicitly with the subject of extraction and the refining of mined materials.[37] For sure, many classical and medieval authors deal with certain aspects of mining, metallurgy, and more generally mined materials (e.g., Pliny, Isidore of Seville, Bartholomeus Anglicus, Thomas of Cantimpré, Vincent de Beauvais, and Albertus Magnus), but only in the late fifteenth and sixteenth centuries do manuscript and print books begin to circulate that take up the specific question of extraction, thus contributing to fashioning mining and extraction into coherent bodies of artisano-scientific knowledge.[38] A brief *tour d'horizon* is not out of place. The sixteenth century opened with Calbus of Freiberg, who, as well as helping "establish a humanist Latin school" in the important mining town in Saxony, publishes a work focused on veins and metal ores titled *Bergbüchlein* (1505).[39] The proliferation of such publications continues with an anonymous book about assaying, the *Probierbüchlein* (c. 1510–1520),[40] and with Vannoccio Biringuccio's *De la pirotechnia* (1540).[41] From the mid-1550s dates the hand-copied and richly illustrated *Schwazer Berguch*, probably the work of Ludwig Lässl, an "extensive compilation of mining law, custom, and regulations."[42] In the 1570s appear Samuel Zimmermann's *Probierbüch* (1573),[43] Lazarus Ercker von Schreckenfels's *Beschreibung der Allerfürnemsten mineralischen ertz und Berckwerks* (1574),[44] and Ciriacus Schreittmann's *Probierbüchlin* (1578).[45] Finally, Modestin Fachs's *Probier Büchlein*, written circa 1569, went to print in 1595.[46] To this list of (mainly) German-language titles, and surely more significant than any of them, are of course the Latin works of another German author, Georgius Agricola, especially his *Bermannus* (1530), a dialogue between two doctors and a mining specialist; his *De animantibus subterraneis* (1549), about creatures that live underground; and most importantly his *De re metallica* (1556), which would soon become the "miner's Bible."[47] The Germanocentricity of this corpus is a historical fact with a clear cause: The German mining industry was, at this time, the most developed and technologically advanced in all of Europe. *Exterranean* is, on the contrary, not particularly Germanocentric and definitely not primarily focused on Germanophone texts.[48]

Such an enumeration of texts conveys the quantity of early modern humanist works about mining—but more significant here is that these works, in different ways, both changed the status of mining as a profession *and* collectively brought mining within the bounds of humanist culture and practice.[49] Indicative of this is how the eponymous mining specialist of Agricola's *Bermannus* combines firsthand experience of mines and mining with knowledge of ancient languages and authors.[50] Indeed, from the opening of the dialogue, Bermannus and his interlocutor are presented as being "as skilled in literature as in minerals [*non minus literarum, quam rei metallicae periti*]."[51] While the dialogue presents itself as an actual walk around a mining area—"Let us head directly to the pit entrance for this vein! [*ad venae istius fodinas ascendamus*]"[52]—and while it focuses on various hands-on technical aspects of mining, it is a fundamentally humanist work in which philology, translation, and interpretation of classical authors is paramount, an emphasis made clear from the opening sentence.[53] Throughout, questions abound concerning nomenclature and correspondence with (or difference from) classical authors, as when it is asked: "This plumbago [i.e., galena, aka black lead], is this not what Pliny calls *molybdaena* [*quam Plinius* μολύβδαιναν *etiam vocat*]?"[54] Yet classical authors are not authorities but interlocutors: "Truth, more powerful than Pliny, will defend us [*veritas ipsa, quae potior Plinio est, nos defendet.*]"[55] The practical and the philological combine in many places, as here: "Let us climb up this mound. This is ochre or, if you prefer speaking Latin, *sil*."[56]

The *homines* who wrote these various sixteenth-century books about mining are not, in any way whatsoever, to be taken as ecological thinkers or as environmental activists. On the whole, they describe *and indeed praise* extraction in the context of an early modern mining boom in which the birth of modern capitalism brings about an increase in the scale of mining operations.[57] As a rule, these writers dismiss—such is the case of Agricola—"those who point to the dangers and unhealthiness of mining."[58] When I read such texts, I will thus frequently be trying to access a sense of mining and extraction that was not at the forefront of the authors' minds or intentions but that shines through nonetheless. *Exterranean* is not, in any case, a history of books about mining. With the exception of Georgius Agricola, who receives extensive treatment in Chapters 3 and 4, most of the aforementioned authors will not be discussed in great detail. I will be reading Agricola as a humanist mining expert, but I will also be reading many non-mining-expert (but equally humanist) authors for the ways in which they can provide access to the exterranean.

ASIDE: THE ORBIS HYPOTHESIS

In attempting to engineer connections between early modern texts and images and the geoecological reality of our collective present under the banner of the Humanist Anthropocene, I am not necessarily propounding (nor refuting, for that matter) that the Anthropocene began in 1610, a claim that, having generated much discussion—and about which I have been asked many times during presentations of the present project—deserves brief review. Simon L. Lewis and Mark A. Maslin advanced such a hypothesis—the "Orbis hypothesis"—in a 2015 article published in *Nature*.[59] Expanding upon Alfred W. Crosby's concept of the Columbian Exchange and drawing on high-resolution Antarctic ice-core records, the two scientists argue that the arrival of Europeans in the New World in 1492 and the subsequent century of slaughter of indigenous populations—whose numbers fell by approximately fifty million—might serve to mark the beginning of the Anthropocene.[60] The hypothesis turns mainly on the fact that the huge number of deaths resulted in a near cessation of farming, a reduction in fire use for land management, the regeneration of over fifty million hectares of forest, savanna, and grassland, and thus in a significant increase in carbon sequestration. Lewis and Maslin thus propose that "the 7–10 ppm dip in atmospheric CO_2 to a low point of 271.8 ppm at 285.2 m depth of the Law Dome ice core, dated 1610 (± 15 yr) [might serve as] an appropriate

GSSP [Global Boundary Stratotype Section and Point] marker."[61] They call this proposal the "Orbis hypothesis" because "post-1492 humans on the two hemispheres were connected, trade became global, and some prominent social scientists refer to this time as the beginning of the modern 'world-system.'"[62] From a scientific standpoint, the date of 1610 thus appears—and as a nonscientist I rely here on Lewis and Maslin—defendable.

The Orbis hypothesis sparked reactions in various quarters, including from scholars in the humanities, two of whom (Dana Luciano and Steve Mentz) it is useful to discuss in this Introduction for their direct bearing on what I mean by the Humanist Anthropocene. Dana Luciano, a specialist of nineteenth-century American literature and culture, called the Orbis hypothesis "compelling" soon after the *Nature* article appeared, especially for the way it makes evident the extent to which what is at stake in the naming of the new epoch is "what kind of story can and should be told about human impact on the planet."[63] From the point of view of narrative, the 1610 date thus means that the origins of the Anthropocene are "not simply human indifference to nature, but human disregard for other human lives," causing Luciano to suggest that we might productively talk of "The Inhuman Anthropocene." In terms of chronology, part of her response to Lewis and Maslin relies on the Jamaican novelist Sylvia Wynter's account of the aftermath of 1492 and argues that what occurs after 1492 is a "spread of a *humanism* that has failed much of humanity," such that the "Inhuman Anthropocene" descends from this "inhuman humanism."[64] As valuable as it might be to talk of an "Inhuman Anthropocene," to speak of an "inhuman humanism" *within the post-1492 context* in which early modern humanism flourishes stricto sensu strikes me as conceptually confusing: Luciano's proposition is based on Wynter's positioning of 1492 as a direct result of a humanism defined as an "increasingly secularized, that is, degodded, mode of 'subjective understanding.'"[65] Colonization was clearly sadistically inhuman— but that does not make humanism a synonym for secularism and cruelty, as I hope the above untranslation exercise suggested.[66] The early modernist Steve Mentz, for his part—and avoiding the term *humanism* altogether—authored an affirmative response to the Orbis hypothesis that dwells not only on its emphasis on inhumanity but also and especially on its inclusion of all the non-human agents that played a role in the colonization of the Americas (smallpox, influenza, malaria, etc.).[67] Mentz also points out that the hypothesis productively erases the "radical newness of today and submerges it back into History, with all of history's messiness and swirl," an erasure and submersion he clearly supports for the interpretive possibilities that they open, for they push in the direction of "beheading Anthropos."[68] Both the sciences and the humanities thus adduce convincing arguments for a 1610 Anthropocene[69]—but acceptance of that dating is not a keystone of the Humanist Anthropocene, which implies *not* (necessarily) a chronological overlapping (in Mentz's words: "It turns out that this latest thing is also an old thing") but a collision between, on the one hand, a body of writings and images (those of early modern humanists) and associated practices (philology, translation, interpretation of classical authors) and, on the other hand, the Anthropocene.

DIALOGUES

The present book clearly entertains deep connections with much recent research and critical thinking across numerous disciplines—more so than I could ever account for here. Theoretically, it develops affinities and engages, directly and indirectly, with various overlapping theoretical traditions, most especially with the French tradition of what Stephanie Posthumus calls *éco-pensée*, represented in particular by the thought of Michel Serres and Bruno Latour,[70] as well as with the thought of Timothy Morton (his rejection of Romantic "Nature," his notions of the "mesh" and the "hyperobject," his "dark ecology") and—for their collective (but sometimes different) elaborations of flat ontology—the work of Jane Bennett (her vibrant materialism), of Quentin Meillassoux (his framing of the problem of correlationism), and of—to a lesser extent—the OOO philosophers Graham Harman, Levi Bryant, and Tristan Garcia.[71] It would be laborious and pedantic to introduce these well-known thinkers in their full complexity here—my engagement with them will be more obviously detectable by some readers than others, a situation I prefer to making the current book too heavy, too beholden to specific lexicons.[72] This book is moreover not the first to put the early modern period in dialogue with questions of ecology and with the Anthropocene: The work of scholars such as Bruce Boehrer, Todd A. Borlik, Gabriel Egan, Ken Hiltner, Steve Mentz, Vin Nardizzi, Karen Raber, Jeffrey Theis, Robert Watson, Tiffany Worth, and many others offer multiple environmental and ecological approaches to early modern literature and culture.[73] Book series such as Amsterdam University Press's Environmental Humanities in Premodern Cultures will likely lead to much more important work in these areas.[74] This large body of work produced by early modernists is rich indeed—yet very little of it deals in any extended way with literature written in languages other than English, especially when that literature has not been translated. This fact is part, of course, of a larger context: As Louisa Mackenzie diagnosed in 2012, ecocriticism, whatever its chronologies, is "largely focused on cultural productions in English."[75] The situation is, to some extent, evolving: Karen Thornber's study of environmental crises in East Asia via the literatures of China, Taiwan, Japan, and Korea, for example, stands out against the English-only tendency of the environmental humanities.[76] A growing number of scholars working in multiple languages and traditions are bringing ecology into dialogue with literature.[77] And the question of comparative ecocriticism is now receiving direct treatment,[78] as is the specific question of how to think about language difference and translation in the age of the Anthropocene.[79] Nonetheless, it is a symptom of the ongoing dominance of English—albeit perhaps a symptom that originated as just a "slip of the keyboard," as one colleague put it to me in conversation—that the MLA's Ecocriticism and Environmental Humanities forum defines itself as "a scholarly practice *within English Studies*," as if the MLA were not the primary professional organization for scholars of all modern literatures.[80] Moreover, despite the growing number of scholars taking up ecology in relation to non-Anglophone literatures, the focus is rarely on pre-Enlightenment literature.[81] In writing *Exterranean*, I have learned a lot from many of the thinkers mentioned here—but I am also trying to nudge the conversation in some new directions: In the chapters that follow, I take up early modern texts written in, and frequently never translated from, Latin as well as various European vernaculars, and I do so with the principal goal of seeing how they can provide a subject position (that of the *homo*) for inflecting our theoretical understandings of the Anthropocene. My goal is primarily to see how this subject position can lend its agency to the undifferentiated *anthropos* of our times.

One book in particular—Jeffrey Jerome Cohen's *Stone: An Ecology of the Inhuman* (2015)—merits particular mention here.[82] I started to present parts of *Exterranean* shortly before Cohen's book appeared and thus before I had read it, and audiences everywhere soon remarked on the proximity of some of our claims.[83] *Stone* is indeed, as reviewers have also noted, a remarkable book. Its chapters ("Geophilia," "Time," "Force," "Soul"), each complemented by its own personal and more tactile "Excursus," reject stone's inertness (after Deleuze and Bennett) and explore its lithic force and vibrant agency, its ability to mediate and to create reciprocal intimacies between the human and the nonhuman (terms usefully collapsed in the book's idea of *in*human ecology),[84] from the conjoined perspectives of contemporary theory and medieval literature, purposefully producing a sense of (productive) "vertigo" in order to situate stone as part of a "cross-ontological alliance."[85] *Stone* and *Exterranean* plainly work at articulating some similar questions from some of the same theoretical footholds. On the one hand, there is a fundamental difference in emphasis: *Exterranean* seeks specifically to discuss not stone but what of the Earth remains in our affiliations with what has been taken from it and what, when it is still connected to the Earth, anticipates its future disconnect. On the other hand, Cohen's thought—in added conversation with that of Tiffany

Werth—will be particularly close at hand in Chapter 5, on "Geomedia," a chapter focused on limestone. Most of my other chapters will also empathize with and seek out Cohen's sense of (always difficult, always uncertain, always somewhat of a "gamble") affiliations between the human and the non/inhuman, but they will not be primarily focused on stone in particular.

Faced with accounting for the numerous conversations and scholarly affiliations with which this book engages, I am reminded of Montaigne's pressing exclamation in his chapter "De la praesumption" ("On Presumption"): "Pour qui escrivez vous?" ("For whom do you write?").[86] Even the simplest answer to that question in my own case—that this book is written for early modernists interested in ecology and for so-called environmental humanists, *éco-penseurs*, and new materialists curious about what early modernity, precisely in its non-modernity, might provide—means that, as a double *passeur* trying to speak in two (or more) directions at once, I run the risk of catching no one's ear and perhaps of making a purely "additive" offering to any given field.[87] My intention and ultimate hope is that, as it brings the *homo* into dialogue with the *anthropos*, and even as it begins with very specific texts and theoretical inheritances, this book will forge a concept (the exterranean) that will transcend specific disciplines and be useful in a more general way.

WHAT LIES AHEAD

The chapters that follow explore humanist-produced texts and images that give meaning to the exterranean. Read as standalone essays, they each offer a certain ecology of extraction, that is, a certain sense of what the extractive process is, of where it begins and ends, and of who and what are involved as human and nonhuman agents. Read in sequence, however, they follow a specific path: Section 1 (Chapters 1–3) examines the exterranean from the point of view of the whole (e.g., Terra, the planet, the globe, the world system, etc.), Section 2 (Chapters 4–5) from the perspective of the sites of extraction (i.e., hillsides, mineshafts, etc.), and Section 3 (Chapters 6–7) from the position of extracted matter (more precisely, limestone and salt). Despite first impressions, the advancement is not primarily a scalar one, that is, from big Terra to small extracted matter. The exterranean cannot be located. It is relational. The very process it defines—taking small bits from a big whole—makes it impossible to select, definitively, one scale over another, for both are always present or possible.[88] As such, the exterranean always involves both big Terra and small extracted matter. What changes from section to section as we go through this book is the part of the exterranean to which the early modern archive—and these readings thereof—strive to give a voice. In other words, the present book is somewhat structured by the very problem of access to its central concept, in order to elaborate that concept from multiple perspectives, to explore its semantic limits.

While this book opens up the humanist archive of early modern Europe, it does so very much with the hope of opening it up to, and for, the twenty-first century. In line with this goal, because much of its inspiration comes from contemporary thought and art, and with the further hope that such connections can make this book visible and relevant to those who might otherwise not have picked it up, each section begins with a short excursus whose purpose is to frame the section's central concerns and questions via brief discussion of contemporary (twenty-first century) artistic creations, either theatrical (Sections 1 and 2) or photographic (Section 3). For the reader familiar with the early modern period, each excursus will further serve to defamiliarize the questions, a productive step, I believe, in being able to approach sometimes familiar texts with fresh eyes.

The three chapters of the first section, "Terra Global Circus," collectively examine the exterranean from the perspective of Terra. The section is titled after a recent experiment in collaboration between science and theater, namely *Gaïa Global Circus*,[89] a play, more specifically a tragicomedy, written by Pierre Daubigny within the context of a working group led by Bruno Latour, Chloé Latour, and Frédérique Aït-Touati, which stages "our relationship with this new character [*ce nouveau personnage*]," that is, Gaia, aka the World System (*le Système-Terre*). Rather than defining the Gaia of James Lovelock, Bruno Latour, and others, the play—as I shall describe in more detail in the section's excursus—depicts the difficulty of imagining this whole of which we are part and with which we live.[90] The three chapters of this first section likewise ask who/what Terra is and what relationship it/she has with the matter extracted from it/her. Chapter 1 offers a reading of a late-fifteenth-century Latin text, Paulus Niavis's *Judicium Jovis* (*Judgment of Jupiter*), which grants Terra legal standing, a right it/she uses to enter Jupiter's court and accuse a miner of matricide. Situating Niavis's Terra in a tradition going from Greek Gaia to Roman Terra, especially Ovid's, up through the twelfth-century Terra of Alain de Lille in his *De planctu naturae*, and finally reaching forward to Anthropocene-era debates over the relationship between humans and the Gaia of Lovelock and Latour, the chapter teases out how the *Judicium Jovis* fashions a sense of the exterranean in which Terra is neither just a body (or globe) nor a vital force (i.e., Earth's systems) but both at the same time, such that the text's final locking together of Terra and human activity serves as an argument against forms of ecological thought that depend upon seeing totalities from afar. Chapter 2 turns to a mid-sixteenth-century poetic text, the "Hymne de l'or" ("Hymn to Gold") by the French author Pierre de Ronsard. I read the poem here—recuperating in particular its "vision" of Terre that generations of critics have written off as a mere aside—as a poem ever conscious about gold's exterranean origins and as a kind of poetic counterpart to nonpoetic texts about mining such as Georgius Agricola's *Bermannus* (1500) and Vannoccio Biringuccio's metallurgical treatise *De la pirotechnia* (1540). After analyzing the "vision" of Terre in some detail, especially the way that Terre is described as always already containing not just gold but *mines*, the chapter explores how Ronsard juxtaposes (in somewhat problematic ways) specific sites of extraction. Chapter 3 shifts the focus firmly to a geographical—as opposed to mythological—Terra and to Terra's (quite literal) early modern post-1492 globalization, as when Sebastian Münster says Terra is known to be "globum magnum & rotundum" (a big and round globe).[91] The chapter explores how the post-1492 Terra of cosmographic writing and globe gores[92]—and which might *seem* to be produced only for a divine *seen-from-nowhere* perspective—collide with contemporary presentations of New World mining on this new global Terra. Taken as a whole, the three chapters of this "Terra Global Circus" confirm, like the play from which the section takes it name, that Terra, if indeed global, is rarely just a globe; that to perceive exterranean extraction as removal of matter not just from the part but from the whole is to develop an awareness for Terra's systemic vitality; and that such perceptions are not given but created and contested.

The second section, "Welcome to Mineland," borrows its name from a festival —"Welcome to Caveland!"—organized by Philippe Quesne at the Théâtre Nanterre-Amandiers in the fall of 2016. As I shall note in more detail in the excursus, the festival offered a variety of plays, films, and other events designed to explore life underground. Again, no one meaning was ascribed. Rather, viewers could enter into a variety of sympathetic relationships—for example, with moles tunneling, with worms talking, or with humans caving—in order to descend perceptually underneath the Earth's surface. In an analogous manner, the two chapters gathered under the heading "Welcome to Mineland" switch our perspective from that of the whole (Terra) to the hillsides and mineshafts where matter—changing ontologically—goes from being (part of) Terra to becoming exterranean. Chapter 4, "Sickly Mountainsides," offers a counter-reading of Georgius Agricola's *De re metallica* (*On Mining*) (1556), the first major printed book about mining. Usually understood within a teleological narrative of technological progress, I shall here argue that Agricola's work also, and despite itself, gives a voice to the landscapes from which matter is extracted. I focus first on Agricola's defense of mining, which, perhaps by trying too hard, communicates a fairly coherent account of the damage done by extraction and of the ways in which that activity interrupts the vitality given by Terra. Chapter 5, "Demonic Mines," shifts attention from the hillsides to the shafts and galleries that miners hollow out underground. Via close reading of another work by Agricola, his earlier *De animantibus subterraneis* (*On Subterranean Creatures*), other sections of *De re metallica*, as well as works by the French writer François Garrault and by Paracelsus, this chapter asks if it is possible to understand belief in mining spirits (otherwise known as *kobaloi*, tommynockers, and by a host of other names) as manifestations of the chemical realities, and medical dangers for humans, connected with extracting matter. From this section, it thus emerges that for early modern humanists extraction of matter *ex terra* was never just a question of human agents yielding extractive and controlling mastery over inanimate hillsides and underground rock faces. If Section 1 uncovered Terra's systemic vitality in the whole, then here we see that same vitality appear, and structure extraction ecologies, at the very places where the exterranean occurs.

The third section, "Hiding in Exterranean Matter," turns from the global and local theaters of extraction to pause at, and linger over, what has been extracted, following the guiding principle that we *anthropoi* are not only Earthbound but also bound-to-live-with-bits-taken-from-the-Earth. This section's title echoes that of Liu Bolin's *Hiding in the City n° 95*, which shows a man (Bolin) standing in front of, and as if stained by and thus somewhat indistinguishable from, a pile of coal. The two chapters of this final section likewise ask how we humans relate to extracted matter and how and how much we are aware of its exterranean origins. Many extracted matters could have received particular treatment here; I purposefully selected two of the less obvious targets within a book about ecology: not coal or oil or even metal but limestone and salt. Chapter 6, "Geomedia," focuses on limestone in the context of the Normandy town of Caen, which is largely built atop and indeed from that stone. Caen is thus at once extraction site and extracted matter. Discussing both buildings within the city and the texts of a local historian, Charles de Bourgueville, which describe them and their destruction at the hands of Protestants during the Wars of Religion, I seek out some of the ways in which humans long for that stone's endurance while also worrying about its fragility. Putting geological and human timelines into dialogue, the chapter thus situates the exterranean at the intersection of extraction, longing, and fear. Chapter 7, "Saline Intimacies," studies another and wholly different exterranean matter, salt. Here, I focus on how authors (both early modern and contemporary) attempt to describe salt by listing what it does, which ultimately points toward its chemical properties, as too does the doing of one particularly salty hero, Rabelais's Pantagruel. I next turn to salt's contested exterraneanity via Cellini's *saliera* and a poem by André Mage de Fiefmelin. The chapter ends by discussing a Neo-Latin poem by Conrad Celtis that imagines salt extraction and human descent into the mines in strikingly similar terms. From these two chapters, it becomes clear that collective perception of the exterranean origins of extracted matter is an oscillation more than a given and that it does not determine in and of itself how humans live with bits-taken-from-the-Earth.

Humans are the greatest (the most extensive) anthroturbators. The deepest subaerial burrowers after humans are aestivating Nile crocodiles, which dig down approximately twelve meters below the Earth's surface. Compare that to human mining: Most mining occurs at depths of hundreds, not tens, of meters—some gold mines in South Africa extended downward as much as four kilometers. It is time to start keeping our anthroturbative effects in mind.

PART I

TERRA GLOBAL CIRCUS

The three chapters of this first section examine the exterranean from the position of Terra (the whole). The goal is not to *define* Terra but to examine the multiple "things" Terra can be—it is misleading to gloss Terra simply as "Earth"—and the various ways in which texts and images produced by early modern humanists perceive the phenomenon of extraction from the point of view of the "whole" from which a "part" is taken. I gather these three chapters together under the heading Terra Global Circus to establish a purposefully nondefinitive and uncertain genealogy between ancient thought, early modern humanism, and contemporary theory as well as to appeal to a certain openness found in theater that is usually denied to academic writing. I adapt this title from *Gaïa Global Circus*, a play created collectively by a number of people, including but not limited to a working group composed of Bruno Latour, Chloé Latour, and Frédérique Aït-Touati, the author Pierre Daubigny, the dramaturge Elsa Blin, and the scenographer Olivier Vallet.[1] The play is impossible to summarize—only fragments can be grasped here.

To get an idea of the play and why I adapt its title, it is useful to note that it is actually the second play to emerge from the aforementioned working group. The first, *Cosmocolosse*, told approximately the same story, about a modern Noah who attempts to build an ark from plastic waste. The characters in *Cosmocolosse* all owned their individual voices and possessed individual characters and identifiable subjectivities.[2] *Gaïa Global Circus*, a rewriting by Daubigny first baptized *Ciao les humains!* (*Bye Bye Humans!*), rejects such clarity, as well as didacticism and—as far as possible—anthropocentrism, in order to register "better the uncertainty (about *who speaks* and *about what*) in which all the [play's] protagonists find themselves when it comes to talking about the Earth."[3]

As the play unfolds, characters appear in Hazmat suits, dressed and performing as UN translators, engaged in aesthetic experiments and (as in a circus)

FIGURE 2. *Gaïa Global Circus.* © Paula Court.

various "acts," and generally searching for linguistic and properly theatrical ways of giving a voice to Gaïa, to nonhumans, and to humans who struggle to comprehend. Gaïa, perhaps most present in the mysterious canopy that hovers over the stage, suspended by huge helium-filled balloons, for most of the show before it is then carried *over* the audience, is thus never *out there*, never totalized, never whole, but always everywhere and *in* everyone and everything. At one point (Figure 2), we see a character silhouetted against the projected image of (something like) water in a Petri dish placed on an OHP, one of the moments that likely led one spectator to propose that the play stages—and induces an experience of—the fact that "we do not live on a 'Blue Marble,'" a (rejected) image of our planet that "symbolizes an objective, holistic, impersonal earth." In place of *that* planet, the play allows Gaïa to speak as a fragmented and contradictory nontotalized totality.[4] At another point, a female character dressed in white says: "Now, I am Gaïa—and you too, asshole, are Gaïa."[5]

The chapters of the following Terra Global Circus substitute for this contemporary Gaïa—of James Lovelock, Lynn Margulis, and Bruno Latour—the Terra as perceived, configured, and constantly reimagined by early modern humanists. They collectively show Terra to be as nonmonolithic, as fragmented, as contradictory, and as urgently disputed as the contemporary Gaia and thus that the ways in which "part" and "whole" relate in the context of extraction is equally contentious. Chapter 1 will offer a reading of Paulus Niavis's Neo-Latin and frankly playlike *Judicium Jovis* (*Judgment of Jupiter*) (c. 1490) that will emphasize how it produces a sense of the exterranean in which Terra is granted standing, in which Terra induces a certain sense of connection between herself and extracted matter, and finally how the text—in its own refusal to provide a didactic solution for exterranean violence—enters into contemporary debates about definitions of Gaia. Chapter 2 turns to a poetic text, Ronsard's "Hymne de l'Or" ("Hymn to Gold"), which also depicts Terra as an animate being and extracted gold as "vibrant matter" but, more, as an exterranean fantasy that obfuscates a truly material sense of extraction. Chapter 3 asks what happens when, in the wake of the 1492 "discovery" of the New World, Terra is globalized. How, in that context, are whole (Terra) and extraction rethought in terms of the specificity of New World mining sites? How does anthropological difference and the colonial system change the sense of the exterranean, and is this anything like the kind of particularization (*this* population versus *that* population) proper to the climate justice movement in the Anthropocene? Discussion will draw here on works by Sebastian Münster, Michel de Montaigne, Bartolomé de las Casas, and others.

CHAPTER 1

TERRA HAS STANDING

Take a look at Figure 3, a woodcut from Paulus Niavis's *Judicium Jovis* (*Judgment of Jupiter*), a work first published in Germany in the 1490s and now more or less forgotten. It summarizes the following narrative. A hermit, out walking (top left) in the Ore Mountains—or Erzgebirge—on the border of Saxony and Bohemia, today's Czech Republic, then the heart of Europe's mining industry, stops to pray (top right). After a short nap (not depicted), he continues wandering through the hills and ends up stumbling into a strange walled garden. He sits down, looking somewhat overwhelmed, and finds himself having arrived at a trial, currently in progress. The king of the gods, Jupiter, sits enthroned (center), holding, Jesus-like, a medieval *globus cruciger*, a symbol of global authority. To Jupiter's left (our right) stands a trim man who carries the tools of his trade, a pickaxe and a hammer. He is a fifteenth-century miner, complete with his recognizable miner's hood and accompanied by his defense team, namely his *penates*, or household deities. To the right of Jupiter (our left) stands a woman. With one hand, she points a finger at the orb in Jupiter's hand, upon which his eyes also focus; with the other, she holds out her dress, so that everyone can observe that it is full of holes. The woman, we learn from the text, is Terra ("mulier, nomine terra").[1] Each of the holes in her dress, created by a miner's pickaxe, materializes the *ex* of the exterranean. As the woodcut also shows, Terra is accompanied by lawyers: the messenger god Mercury; Bacchus, the god of wine; and Ceres, the goddess of agriculture. In the trial that follows, Terra accuses the miner of causing the exterranean holes in her dress. The confrontation thus puts extraction on trial, not as a local problem, not as a situation that affects only the *part*, that is, the Ore Mountains, but as one that affects and receives a response from the whole, Terra. But what kind of whole is this Terra? And what kind of part-whole relationship does the *Judicium Jovis* problematize? How does Terra conceptualize the exterranean?

FIGURE 3. "Terra versus the Miner." Woodcut from Paulus Niavis, *Judicium Jovis*. Bayerische Staatsbibliothek, Munich. Call number: 4 Inc.s.a. 1334. urn:nbn:de: bvb:12-bsb00030187-4.

SQUALOR AND ENTRAILS

Speaking for the plaintiff, Mercury addresses Jupiter as follows: "I bring before you [news of] great sorrows, huge wrongdoing that should not be tolerated, as well as [news of] incredible and unjust suffering."[2] Using a bodily lexicon, Mercury explains how Terra has been hurt and betrayed (*laesam*), stained with her own blood (*cruentam*), punctured (*perforatam*), and wounded (*vulneratam*),[3] concluding that the miner has committed matricide by attacking his mother's body (*corpus, ventrem*), by digging deep into her inner organs, entrails, and members (*viscera, intestina, membrorum*), a discourse made active by a plethora of verbs of violence (*perforat, laedit, offendit, debilitate,* etc.).[4] Terra's Pagan lawyer depicts the miners as relentless in their extraction, working day and night, never sleeping in their boundless desire to deplete Terra.[5] These miners, argues Mercury, are strange humans indeed: They "flee the nourishing light of the sky [*almum fugiunt coeli lumen*]" in preference for "the dark foulness of the Earth [*atrum . . . terrae squalorem*]."[6] Exterranean activity is thus presented as resulting from a human obsession with the inside of a maternal body and its somber *squalor.* Various technical matters are addressed; then Mercury gets even more personal and focuses on emotion: Does the miner feel no sorrow (*cordolium*)? He encourages him to look at his mother's ruined body, the fissures (*rimas illas*) that he has wrought in her, the blood, her pale face, leading up to a phrase that captures the full force of exterranean violence: "You strike out at your mother . . . and you attempt to shatter her entrails" ("Tu contra matrem laedis [et] interiora eius disrumpere conaris").[7] Mercury thus establishes both the facts, that is, what the miner did, *and* he furnishes a specific affective register by replaying (as if video evidence) the specifically matricidal nature of the violence exacted on Terra.

By depicting Terra as Terra Mater, Niavis's Mercury is, of course, building on a very long tradition. As obvious as this connection may at first appear, the implications of Terra's motherliness, especially as regards what kind of whole she is to the parts that get extracted, are not quite so self-evident. When the Homeric Hymn evokes Gaia as a "universal mother [παμμήτειραν]" who "nourishes everything there is on the land . . . and in the sea and all that flies,"[8] Gaia is motherly because of her capacity to nourish, that is, to be vital and to pass on vitality to other beings. Gaia is foremost a divinity of generation and production: "He is fortunate whom your heart favors and privileges, and everything is his in abundance."[9] The Orphic Hymn similarly addresses Gaia as the "mother of men / and of the blessed gods," as she who "nourish[es] all" and who brings all "to fruition" (and who also has the power to "destroy all").[10] Both Greek hymns depict Gaia "more as a natural force than [as] a personified entity," thus forging close affinities between Gaia and ancient poetic representations of *physis* (φύσις)—not to be confused with the Romantic Nature that Timothy Morton calls for us to do away with.[11] In other words, Gaia is always a motherly universal vitality but only sometimes a (personified and complex) body. The same will be true for Roman Terra well into the Middle Ages and beyond, as in the anonymous *Precatio Terrae* and *Invocatio omnium herbarum.*[12] The anonymous *Precatio*—conserved in a number of codices, the earliest of which dates from the sixth century[13]—

addresses Terra as "Goddess revered, O Earth, of all nature Mother" ("Dea sancta Tellus, rerum naturae parens"), noting further how she alone supplies "each species with living force" ("sola praestas gentibus vitalia")[14]—as with Gaia, Terra is here more immanent vitality than a penetrable and minable body.

It is clearly from one specific perception of Roman Terra that Niavis inherits the idea of exterranean activity as matricidal. Essential here is Pliny: When he talks about the mining of metals in his *Natural History*, he describes extraction as a process that involves penetrating the "entrails" (*viscera*) of "our sacred parent" (*sacrae parentis*), granting to Terra at one and the same time the status of maternal divinity *and* a three-dimensional body that seemingly makes her spatially coextensive with Earth.[15] Terra's bodiliness—and the fact that extracting matter from Terra might be matricidal—are further textured when Pliny adds that Terra provides for human needs *on her surfaces*. With that humans should be content, says Pliny: "How innocent, how happy, how truly delightful even would life be, if we were to desire nothing but what is to be found upon the face of the earth [*supra terras*]."[16] Extraction *ex Terra* rather than living *supra terras* leads in the *Natural History* to "ruin," "avarice," and the "exhausting" of Terra. Taking an ecofeminist perspective, Carolyn Merchant sees Pliny's description of Terra as being part of a wider Roman tradition that, by casting Terra as mother, serves as an "ethical constraint" and a "restraining force" vis-à-vis extractivist mentalities: *Terra Is Your Mother, Do Not Abuse Her!*[17] This conception of Terra gives her a maternal personified body, which would frequently be taken up in sculpture and painting (Figure 4), while also—for example, in Pliny's discussion of extraction—making Terra's body that of the Earth. The coextensivity of the two bodies is an open question and one that pertains in the *Judicium*. Who/what exactly accuses the miner?

It is thus primarily a Terra inherited from Pliny and other Roman authors that Niavis draws from when he has Mercury accuse the fifteenth-century miner of matricide. The miner, as we might expect, says he is innocent. Speaking on behalf of himself and his fellow miners, he argues: "We truly do not attack our mother, we do not hurt her" ("Non enim matrem offendimus, non eam vulneramus"), adding for good measure: "We do not hurt any god" ("nullum laedimus Deum").[18] He next rebuts by passing the buck: "We are fulfilling our duty" ("officium nostrum explevimus"), explaining that the order to extract came from Jupiter.[19] So far, the miner's argument is flimsy at best. His final point, however, cuts to the very heart of Terra's motherliness. As if in direct response to Pliny, the miner says that because she hides (*abdit, occultat*) those

FIGURE 4. Terra on the Ara Pacis, Rome. © Phillip John Usher.

metals so essential to the forging of coins and thus so necessary for facilitating trade between regions, this so-called Terra Mater should really be referred to not as mother ("genetrix") but as wicked stepmother ("noverca").[20] To make sure the point is remembered, the miner repeats his use of *noverca* several pages later, acknowledging that, indeed, the miners are removing matter from Terra—but that it is from a stepmother's entrails, not from a mother's.[21] In a later direct confrontation between Terra and the miner, Terra responds, presenting herself as a pious mother ("pia mater") who suffers from having a hardened and obstinate son ("induratus es, nate . . . obstinatus").[22] The miner thus attempts, by refusing his mother her biological maternity, to change the reigning affect and to silence and reject the idea that Terra (the planet) and Terra Mater (the vitality-granting universal mother) are inherently and multiply interconnected.

VIOLENCE ON A BODY

The fact that Niavis builds on the Roman notion of Terra as maternal body—and thus of exterranean activity as matricidal—does not resolve the question of *what* kind of body Terra *is*, nor the whole-part problem. It is tempting to think about Terra's accusing of the miner in the light of Michel Serres's discussion, in his *Natural Contract*, of how ecological thought rematerializes the Earth: "The whole world is present as an object [*le monde entier s'objective*]," throwing the individual "outside [*dehors*]" and casting her "off from the globe."[23] Niavis's miner, likewise, had been standing on Terra, mining Terra, but seeing only the topographical features of the Erzgebirge, when all of a sudden, he is forced to reckon with this Terra. If this *is* however what happens, how do we reconcile the fact that as this happens, the miner is clearly *still* standing somewhere? If Terra is standing there as a woman/a mother, then on what/whom do she and the miner stand? What kind of doubling is this? To advance, let us examine how Niavis describes the body of Terra.

First referred to as "a woman, named Terra [*mulier, nomine terra*]," Terra is further qualified as being "noble [*honesta*]" and "freeborn [*liberalis*]." Of a certain age ("gressum maturius attingens"), she wears a green robe ("veste viridi amicta"), a detail obviously not visible in the woodcut (Figure 3). Her eyes are said to gush with tears like a fountain, something also not visible in the woodcut. Her clothing is torn ("vestitum distractum"), and her stomach, too, is punctured in many places ("ventrem denique vidisses multimode perforatum").[24] The woodcut does not show the punctured body, only the exterranean dress that stands in as a metonymy thereof. One feature of the woodcut is not referenced in the text. In the woodcut, Terra wears a kind of head covering known as *bianche bende*, or "white fillets," typical of early modern widows—further suggesting, then, her age and state of abandonment.[25] As is already evident, Niavis's Terra stands out against and takes meaning from a whole host of related depictions, classical, medieval, and early modern, of Terra—but also of allied figures, especially Natura. A full archaeology is not possible here, but one figure imposes itself as a likely precedent.

The figure of Terra in the *Judicium* shares many traits with that of Natura in Alain de Lille's thirteenth-century *De planctu naturae* (*The Complaint of Nature*), a text central to a significant medieval shift pertinent here. Whereas classical authors "occasionally personified nature" but usually only "in passing," the twelfth century witnessed a turn toward "far more extensive and elaborate fictions" that included Natura as a central character, as in Alain's text.[26] Here, Natura arrives in the "lowlands of this mortal earth," crying and lamenting, in order to confront the "accursed vices of men" and to excommunicate the guilty parties.[27] Alain's Natura is very bodily; she is described as having hair "which shone not with borrowed light but with its own," "twin tresses [that flow] loosely"; she wears a richly decorated diadem; she is dressed in a "garment, woven from silky wool and covered with many colors" and on which "as a picture fancied to sight, was being held a parliament of the living creation" (a hawk, a kite, a falcon, an ostrich, a peacock, etc.). She is further attired in a fine mantle ("its white shaded into green") whose folds ("the color of water") sport images of various sea creatures (a sea-dog, a sturgeon, a herring, a mullet, a salmon, etc.)—"These pictures, finely drawn on the mantle in the manner of sculpture, seemed by miracle to swim." Alain's Natura—as well as Elizabeth I's dress most likely based on it in her Hardwick Hall portrait (c. 1599) (Figure 5)[28]—shows animal and plant beings as metonymies for *all* of Nature.

Perhaps most significant here is that Alain's Natura is further described as wearing a memorable "damask tunic" made of "thicker material approaching the appearance of the terrestrial element [*in grossiorem materiam conglobata, in terrestris elementi speciem aspirabat*]," on which are represented the animals of the earth (an elephant, a camel, a gazelle, a bull, a horse, an ass, a lion, a unicorn, etc.)—*this* part of Natura's clothes is, like that of Niavis's Terra, not in pristine condition: "Here the tunic had undergone a rending of its parts, and showed abuses and injuries" ("In qua parte, tunica suarum partium passa dissidium, suarum injuriarum contumelias demonstrabat").[29] Niavis will use different words (*distractum, perforatum*), but his Terra is nonetheless in a strikingly similar position to Alain's Natura. Like Terra in the *Judicium*, Alain's Natura accuses men of abusing their mother: "Many men [*plerique homines*] have taken arms against their mother [*in suam matrem*] in evil and violence" and—specifically—have put holes in her clothes: "[They] themselves tear apart my garments piece by piece [*mea sibi particulatim vestimenta diripiunt*]."[30] The attack of Natura's clothing is, however, wrought in different terms: She is said to be "stripped of dress [*me vestibus orphanantam*]," thrown into shame. To collapse the complaint fully into the ripped clothing, Natura concludes as follows: "This tunic, then, is made with this rent [*per hanc scis-*

FIGURE 5. Studio of Nicholas Hilliard, *Queen Elizabeth I* (c. 1599), known as the Hardwick Hall portrait. National Trust Photo Library/ Art Resource, NY.

suram], since by the unlawful assaults of man alone the garments of my modesty suffer disgrace and division [mea pudoris ornamenta scissionis contumelias patiuntur]."

In the De planctu naturae, Natura is a clothed figure whose clothes have been torn somewhat like Terra's in the Judicium. But the differences between the two figures and texts are just as important, differences that help us see more clearly what sense of the exterranean Niavis's Terra produces. The holes in Natura's clothes are more metaphors for a whole host of human moral shortcomings—holes in the moral fabric, so to speak—than metonymies for actual material violence to the surface depths of the planet. They are precisely not exterranean holes. Alain de Lille's Natura complains of having been attacked in her sense of modesty, and her grievances vis-à-vis humans concern vices such as drunkenness and flattery: "The evidence of evil committees tells . . . fully . . . of the shipwreck of the human race."[31] The only rather indirect allusion in the De planctu naturae to exterranean violence specifically is equally wrapped up in morality, for example, when Natura complains of humans' "cursed hunger for gold [fames auri]" that "dissolves friendships, begets hate, incites anger, sows strife, nourishes dissension," and generally undoes peace and community.[32] Katherine Park, in her study of the figure of Natura in early modern Europe (with particular reference to Italy), has suggested that Natura's representation shifts in the early modern period to emphasize "more strongly . . . the working of matter," thus separating matter "from the realm of morals."[33] Alain's text clearly belongs to the moment before the shift that Park identifies occurred, while Niavis's Terra shares with Park's early modern Natura this new emphasis on the material rather than the moral.

AGRICULTURE VERSUS EXTRACTION

Other aspects of Niavis's text further develop the specificity of Terra's complaint about exterranean activity. The fact that Niavis's Terra is accompanied—and defended—by Bacchus and Ceres reinforces both this insistence on the materiality of exterranean activity (as opposed to moral violence, as in Alain de Lille) and her affinity with Gaia-as-vital-force. Bacchus focuses his speech on the direct relationship between the production of exterranean materials—"[the miners] dig, disturb, and annihilate . . . the hard rocks [Fodiunt, perturbant, annihilant . . . dusirrimos rupes]" as they "search for metals [quaerunt metalla]"—and the damage that is done to the Earth in general—"they tear into the peaks of mountains [distrahere montium cacumina]"—and to his vines in particular, the latter being his central complaint: The miners mangle the vines (vineas lacerare).[34] In other words, taking matter ex Terra means interrupting its vitality and thus, as direct consequence, his grape production—this is both a local phenomenon (the vines) and a bigger one (Terra's vitality that is infused in the plants). Ceres's argument is similar: Staying wholly true to the ultimate Indo-European root of her name (which means *to satiate, to feed*), she performs her role as goddess of agriculture, crops, and fertility by making an argument against mining based on the fact that exterranean products such as gold and silver cannot be eaten and digested by the human body ("indigestibile aurem est, et aurum et argentum")—neither can they satiate ("non pellit") hunger or thirst.[35] Instead of mining, Ceres argues, the fields should be cultivated ("agros colere") in order to support and nourish human life ("humama vita sustentatur").[36] Bacchus and Ceres speak *for themselves*, but as Terra's lawyers they also speak for that part of Terra that is not planet per se but planetary vitality.

The opposition between taking matter *ex Terra* and agriculture finds its origin in Ovid's *Metamorphoses*. In Book I, Ovid tells a story of transformation that is generally assumed to function as the chronology of a fall, that is, from a vice-free and harmonious "golden age" to a corrupt and war-beset "iron age," but that also performs—to take up an idea from Dipesh Chakrabarty and Bruno Latour—a collapsing of (human) history and geohistory.[37] Each of Ovid's four ages espouses a different form of terraformation, as a quick rereading demonstrates. In the first, golden, age the Earth "without compulsion, untouched [*intacta*] by hoe or plowshare, of herself gave all things needful," such that humans, "content with food which came with no one's seeking, gathered from the mountain-sides, cornel-cherries, berries hanging thick upon the prickly bramble, and acorns fallen from the spreading tree of Jove."[38] In this first age, the Earth is bountifully fertile with no human intervention. The second, silver, age witnesses the invention of agriculture: "Then first the seeds of grain were planted in long furrows, and bullocks groaned beneath the heavy yoke."[39] Farmers now farm the Earth, moreover with the help of animals. The third age is bronze, "of sterner disposition" in which men are "more ready to fly to arms, but not yet impious [*non scelerata tamen*]."[40] I note here that we might also translate this as "not yet polluted"—*scelerare* meaning "to defile," "to pollute." The fourth and final iron age is the age of massive terraformation, for it coincides with the invention of mining:

> The ground, which had hitherto been a common possession like the sunlight and the air, the careful surveyor [*mensor*, the "measurer"] now marked out with a long-drawn boundary-line. Not only did men demand of the bounteous fields the crops and sustenance they owed, but they delved as well into the very bowels of the earth [*in viscera terrae*], and the wealth

which the creator had hidden away and buried deep amidst the very Stygian shades, was brought to light, wealth that pricks men on to crime.[41]

Age by age, humans thus become increasingly significant terraformers: From living off berries, humans progress to grinding furrows into the Earth, for agriculture, and eventually they start to mine. The final iron age is defined by mining and by the unleashing of criminality. Ovid's Iron Age is, in this respect, antiquity's Anthropocene. We see that when Bacchus and Ceres speak out to emphasize the impossibility of agriculture and extraction coexisting because of their antagonistic forms of terraformation, they argue not just for the local "right to cultivate" but, via Ovid, for a more general and large-scale harmonious relationship with Terra. The miner, for his part, clearly understands this, for he responds by trying to direct attention both to the local and to a different relationship of part and whole: He makes the argument that the existence of mining regions is part of a global planetary geography according to which each region is destined to have different material properties and thus to serve a different purpose. A given region might be marked for agricultural production (*una frumentis abundat*—later on, crops referred to include *siligo* [winter wheat], *triticum* [a kind of what], and *frumenta* [corn]),[42] while another is perfect for viticulture (*alia vino*), and a third (watery) one produces fish (*haec piscium generat multitudinem*), such that—as per Jupiter's ordinance, says the miner—exchange between regions is essential, an exchange made easier (*commodius*) thanks to most noble metallic money (*nobilissimum metallum*).

The miner's team of lawyers—his *penates*—repeat this general line of reasoning. In response to Bacchus's point, they argue that the hills in question, the Erzgebirge, are (1) not suited for viticulture: "The land is not fertile [*solum nihil fertilitatis continet*]"[43] but (2) perfect for the mining of metal, which can be used to make money, the most useful thing of all. The *penates* ask a rhetorical question—"what could be more fruitful/useful [*frugalius*] to human affairs . . . than money [*pecuniam*]?"—using the term *frugalius*, derived from *frux* (meaning *produce, crop, fruit, result*), casting metal, as an exterranean matter, almost as an agricultural product. Their response to Ceres is similar: Proper global shepherding (*haec globum terrarum tuendum*), they argue, depends on the coexistence of multiple activities (*variis tractationibus*), agriculture (*sulcus arare*), sailing the seas (*maria navigare*), as well as the encouragement of trade (*studere mercaturae*), and hence mining, metal, and money.[44] Switching registers slightly, the miner's defense team makes a more general defense of metal mining, responding to Ceres's point that gold is useless, by invoking how gold *directly* contributes to the health of all humanity, alluding inter alia to the idea of soluble gold.[45]

Just as Niavis's Terra is like but unlike Alain de Lille's Natura—the former emphasizing exterranean as opposed to moral violence—so the *Judicium*'s appropriation of the Ovidian opposition between agriculture and extraction is more specifically focused on the exterranean than many medieval and early modern representations. Most editions or commentaries of Ovid's *Metamorphoses* that circulated around the time that Niavis was writing belonged to a Christianizing and moralizing tradition best known in the form of the *Ovide moralisé*.[46] A glance at any number of contemporary editions shows just how much room was given to allegorizing interpretation.[47] Giovanni Bonsignori's 1497 Italian translation, for example, includes an interpretation (*alegoria*) that compares Ovid's four ages to the four seasons in any given year, with the iron age being winter—again taking emphasis off the specifically exterranean aspect of the iron age.[48] Or we might look to a 1532 French adaptation of the *Metamorphoses* that summarizes the iron age as being "the driver and guide of all ills," one of which is that the "earth [*la terre*], which until that point had been whole [*entiere*] and shared [*commune*], began to be torn apart [*deschiquettee*] and parcellated [*mises en pieces*]," another of which was the beginning of mining, described as "going into the stomach and entrails of the earth [*aller au ventre & entrailles de la terre*]," causing her to bleed (*saigner*).[49] But while the text is close to Ovid and indeed close to Niavis, the accompanying woodcut demonstrates quite a different emphasis. What we see there is not Terra, and not her violated and bloody body, but rather the concomitant general spreading of bad behavior—via the image of a tavern, in which men give in to their various vices.[50] Compared to these readings, Niavis is somewhat of a literalist, an antiallegorist[51]—his emphasis is clearly on what happens to Terra's surface depths and overall vitality, not on Christian morals or general human corruption.

LOOPS AND WHEELS

Terra reveals herself to be several: As mother, she possesses a body with entrails; as mother, she also infuses all things with vitality. On both counts, extracting *ex Terra* encounters ethical constraints: Humans should not violate their mother, and they should not terraform in ways that destroy her vitality. However, the strangest and most provocative thing about the *Judicium Jovis* is that, and in striking disagreement with the work's title, Jupiter appears, at first reading, never to offer any final judgment. The ethical constraints are *not* upheld in any definitive and tidy way. After hearing the points of view of Terra, the miner, and their respective advocates, Jupiter decides to maintain a neutral position ("ne in partem aliquam inclinari videretur") and to appeal elsewhere for an answer.[52] More specifically, Jupiter elects to consult Fortune via a particularly humanist form of scholarly exchange: He writes her a letter. Her judgment, made via return letter, is long, winding, and somewhat abstruse: Men, says Fortune, should mine and dig the mountains ("homines debere montes transfodere"). They should also farm the fields ("agros colere") and eagerly engage in trade ("studere mercaturae"). They should not hesitate to injure the Earth ("terramque offendere"), jettison science/knowledge ("scientiam abiicere"), or disturb Pluto's realm ("Plutonem inquietare"). Men should also search out veins of metal in rivers ("in rivulis aquarum venas metallic inquirere"), and—and here the long judgment takes a curious turn—their bodies should be swallowed up by the Earth ("corpus vero eius a terra conglutiri"), suffocated by vapors ("per vapores suffocari"), intoxicated by wine ("vino inebriari"), and punished with hunger ("fame subiiei"), all the time remaining ignorant of the best ("quod optimum fit ignorare").[53]

Readers of the *Judicium Jovis* have struggled to make sense of Fortune's sinuous pronouncement, although they generally agree on its central thrust. "Man and the earth are destined to engage in perpetual conflict," says one;[54] humans and nature "engage in a costly contest of mutual attrition," says another.[55] But what we make of that judgment is a more complicated question, especially given both the historical and theoretical complexity of Fortune. Whatever the period we consider, Fortune (in Latin, Fortuna) is far from being a simple personification or goddess of caprice or chance—as one study framed the question recently, Fortune/fortune is "an empty form [*une forme vide*]" whose content is malleable and whose usages are multiple.[56] As we read Niavis, we should keep in mind the archaic goddess of the Roman countryside, "a bearer of abundance, a nurturing figure," her assimilation to Greek Tyche, and especially the complex role of Fortune in Boethius's sixth-century *Consolation of Philosophy*.[57] In the latter text, which remained immensely important throughout the Middle Ages and early modern period and whose first printed edition dates from 1474 (preceding by some twenty years the publication of the *Judicium Jovis*), Lady Philosophy explains to Boethius (who is in prison) not only that true happiness is to be found only beyond the terrestrial and the transitory but also that Fortune's disorder is only apparent—behind it there is God's stability, hence the constant inconsistency of Fortune's Wheel: Atop it, one falls; below, one rises (Figure 6).[58]

We should further keep in mind that historians of the early modern period—starting with a seminal article by Aby Warburg—have generally identified the appearance of a decidedly *new* Fortuna in the late fifteenth and early sixteenth centuries, somewhat conflated with Occasio (and a new kind understanding of time) and with Venus.[59] In passing from the medieval to the

FIGURE 6. "Wheel of Fortune," from Boethius, *Consolation de la philosophie*, Institut de France, Ms.364 f. 9r. RMN-Grand Palais / Art Resource, NY.

early modern F/fortune, we also see a change in iconography: The frequently blindfolded figure who turns a wheel is now often replaced by a usually nude and seductive woman who stands atop a globe, her hair flowing in the wind. It is impossible and ultimately unproductive to link Niavis's Fortune definitively either to the wheel-turning medieval figure or to the newer globe-mounted one, although recent historical research of this change would tend to suggest the former.[60] Most importantly, Jupiter appears to appeal to Fortune because Fortune reigns—as made most clear in Boethius—over the terrestrial realm and that realm's seductive charms: precious metals and other exterranean matters. In the fourteenth century, Petrarch had made Fortune's connection to mining clear in his collection of Latin dialogues, the *De remediis utriusque fortunae* (*Remedies for Fortune Fair and Foul*).[61]

In the fifty-fourth of Petrarch's dialogues, titled "Of finding a golde mine" in Thomas Twayne's 1579 English translation, we hear a debate between Joy, happy to have located this source of riches, and Reason, who is keen to point out the difficulty of extracting gold and the unlikelihood of ever reaping true benefits from that activity. From the beginning, Reason's position is clear: "This hope of ryches, hath been cause of pouertie unto many, and of destruction not unto fewe." Miners "toyle" and "lyfe in darkeness" but turn "little profite." Joy and Reason disagree fundamentally about the nature of fortune (or perhaps *occasio* or propitious opportunity): While the first asserts with confidence that "Chaunce hath offered unto me a gold Myne," the second is of the opinion that extraction will lead to living "more unfortunately." The conclusion is repeated toward the end of the dialogue: Joy argues with clear conviction "I digge earth that will yeeld gold," to which Reason responds, "The travel is certain, but the euent doubtful [*labor certus, eventus ambiguus*]."[62] Pursuing the same humanist interpretive strategy as we have already seen at work in the *Judicium Jovis*, Petrarch cites Ovid's discussion of mining as the event that gives rise to the arrival of the evil Iron Age: "*Men haue entred into the bowels of the earth*," and the exterranean matter leads "*vnto al mischief.*"[63]

A 1532 German edition of the *De remediis utriusque fortunae* provides a fabulous woodcut of miners at work underground, which brings out the sense of uncertainty in extractive activities.[64] For sure, the scene features neither Joy nor Reason, just numerous miners complete with pickaxes and recognizable outfits, as well as one differently dressed man (probably the mine's manager or surveyor), but it is a busy scene indeed that expresses a sense of chaos reinforced by the swirling layers of hatching. We cannot know if Niavis had read Petrarch—it is materially possible but not verifiable. Their juxtaposition here nonetheless helps us appreciate the weight of Jupiter's handing over judgment to Fortune in the *Judicium*. To say, as one commentator has done, that there is in fact "no final judgment" is clearly to miss the point that the production of exterranean matter was—in Fortune's verdict—not only risky business for the miner but a human buying into the riches of a terrestrial realm whose workings are both part of a divine order and wholly unknowable by humans.[65]

In Niavis, Terra is figured as a woman with a body, but her bodiliness is complicated by the fact that her body does not and cannot coextend with Earth as such. She appears in the courtroom in the mountains *on Earth*. Yet she points to the holes in her dress, which *are* holes in the Earth. We could say—that would be the safe answer—that she is *just* a personification. The more compelling (to me) thought is that as we look at Niavis's transposition of Ovidian terraformation, we are looking at a more queer set of intimacies that escape the easy establishment of corporeal boundaries.[66] A "body" is perforated, and because it is Terra that body is terraformed, yet Terra is still visible before us in the mountains in the background. Niavis makes it clear that we are dealing with *actual* terraformation as opposed to terraformation-as-allegory—yet Terra is clearly both planet *and* what inheres within all *physis*. Fortuna's judgment confirms the human place within this: Niavis does not simply replay the Earth-as-mother notion emphasized by the binary reading of Merchant; rather, he suggests that humans will in any case mine *ex terra* then return *a Terra*. Most significant in such an understanding of terraformation is how close this brings us to, and how this offers a figuration of, Bruno Latour's point in *Face à Gaïa* that properly ecological thought *cannot* function with an idea of the planetary that requires the figure of the globe—*to look at the planet as a globe is always to take oneself for God*. Rather, it requires an impossible-to-figure understanding of Earth's systems, in which feedback loops are the main connective.

The *Judicium Jovis* is clearly an ancestor of Anthropocene-era debates about humanity's fate and responsibility vis-à-vis Gaia. The text's relevance, however, is arguably not in its pastness but in its ability to intervene *now* by guiding us through various positions that—more than before—seem opposed. The trial described by Niavis gives form to and endorses a sense of theological abandonment, bringing into the same arena Félix Guattari's averment that ecological thought is "atheist in the best sense," in that it summons a "positive disbelief in God, concerned only with, and respectful of, *terrestrial* life," and Bruno Latour's point that anyone "who looks at the Earth as a globe" necessarily occupies a divine position.[67] As it calls the miner to abandon his localism, the *Judicium* also amplifies Guattari's and Latour's calls for earth over Earth, bringing out—via the impossibility of defining Terra as *either* a totality *or* a vital force—that there is no contradiction in seeking out the planetary in a nontotalized sense of the whole. Rather than parsing "sense of place" and "sense of planet" as Ursula Heise proposes, the *Judicum* makes the case for focusing on the difficulty of such a separation *even as localism is shunned*.[68] The *Judicium Jovis* also provides a way of figuring what comes *after* abdication, that is, a letter and an appeal to the consistent inconsistency of Fortuna. The latter's turning—from *ex Terra* to *a Terra*—entangles humanity into the e/Earth, splintering agency, pleading not for the metaphor of Gaia operating like "the thermostat of a kitchen oven" (Lovelock's early analogy) but for second-order cybernetics (that is, autonomy, not machinic control).[69] Niavis thus speaks for a grounding in *physis* at a planetary level and is thus at odds with the way that Heise locates her notion of a "sense of planet" as something "less territorial."[70] On the contrary, Niavis seems to say, it must be simultaneously *more territorial* and *more terreanized*.

CHAPTER 2

TERRE'S BRILLIANT MINES

The 624 capacious alexandrines of the poet Pierre de Ronsard's "Hymn to Gold" ("Hymne de l'or") (1555) stage quite a different Terra Global Circus from the one discussed in the previous chapter. Being a hymn (< Gk. ὕμνος)—and *not* a trial aimed at reaching a *judicium*—the poem unsurprisingly praises both gold in particular and wealth more generally, overturning a common humanist *topos* (gold is bad).[1] Anyone willing to praise poverty instead, the poet says, should be ready "also to praise the plague" ("loüe[r] aussi la Peste").[2] As one recent reader put it, the poem demonstrates the "unparalleled prestige" of gold in order to offer the reader a "surer moral pathway for admiring riches."[3] The present chapter will nuance existing readings by arguing that the "Hymne de l'or," which draws on many of the same elements and traditions that we have already seen in Niavis's *Judicium Jovis*—although in new combinations and to different effect—should be read not only as a poem that extols the most precious of metals but also as a poem that is ever conscious about that matter's exterranean origins. In many ways—and such an affirmation goes contrary to the general tide of Ronsard criticism—one might say that Ronsard's text is something of a poetic counterpart to Georgius Agricola's *Bermannus* (1500), whose dialogue's eponymous main character, a mining investor, asserts in the most upfront of tones that "almost all of us are hungry for money [*pecuniam avidi*] and want to get rich [*ditescere*] with the least expense and smallest amount of work possible,"[4] adding several pages later that it is precisely the hope (*spes*) for riches that leads the miner "into the bowels of the Earth" ("ad intima terrae viscera").[5] As we shall see, both the *Bermannus* and the "Hymne de l'or" connect gold's value to its exterraneity. The "Hymne de l'or" is also of a piece with the first chapter of Vannoccio Biringuccio's metallurgical treatise *De la pirotechnia* (1540), "Della minera dell'oro, et sue qualità in particolare" ("Concerning the Ore of Gold and Its Qualities in Detail"), which deals with gold and gold mining in ways that combine moral and material registers.[6] As in Chapter 1, which plotted in the *Judicium Jovis* the complex entanglement of Terra and the fifteenth-century Erzgebirge in order to get *one* sense of exterranean ecologies, central to my investigation of Ronsard's poem will be its back-and-forth between whole and part, that is, between, on the one hand, Terra (here called *Terre*) and her allies (especially Nature) and, on the other hand, specific sites of extraction. Ronsard makes no mention of the Erzgebirge, proposing quite different parts to Terra's whole, resulting in an alternative sense of the exterranean.

A VISION OF TERRE

The first extensive mention of Terre/terre in the "Hymne de l'or" is as follows:

> Plus la terre aujourdhuy ne produist de son gré
> Le miel pour nourrir l'homme, & du chesne sacré
> [Lors que nous avons fain] les glandz ne nous secourent,
> Et plus de vin & laict les rivieres ne courent,
> Il faut à coup de soc, & de coutres trenchans
> Deux ou trois fois l'année importuner les champs,
> Il faut planter, enter, prouvigner à la ligne
> Sur le sommet des montz la dispenseuse vigne,
> Tout couste de l'argent, il faut achetter bœufz,
> Pelles, serpes, rateaux . . .[7]

> (The earth today no longer willingly produces
> The honey that nourishes men, and the acorns
> Of the sacred oak, when we are hungry, offer no help,
> And rivers run not with wine and milk.
> Rather, we must with plowshare and sharp coulters
> Torment the fields two or three times per year;
> We must plant and graft and set in lines
> Upon the mount's summit the costly vines.
> Everything costs money. You have to buy oxen,
> Spades, brush hooks, rakes . . .)

Unlike the Earth of Ovid's golden age, Ronsard's *Terre/terre*—with a capital, Terra/Earth, without earth/soil—no longer yields crops voluntarily. She/it must be cajoled by the tools and practices of agriculture, bringing us not only into the silver age but, because agriculture requires exterranean metals, into the iron age, at least if we follow Ovidian chronologies. Ronsard, as we shall see, does not follow such chronologies. Immediately following these lines on *Terre/terre*, Ronsard turns to Nature herself to restate the case and, just as we saw in Chapter 1 how Niavis's Terra shares certain traits with Alain de Lille's Natura, so here *Terre/terre* is/are intimately connected with Nature:

> la Nature ainsi qu'une autre beste
> N'a point l'homme habillé du pied jusqu'à la teste:
> On voit chevaux, lyons, ours, brebis & taureaux,
> Chiens, chatz, sangliers, & cerfz, vestuz de grosses peaux
> Qui defendant leurs corps de chaut & de froidure:
> Mais d'une simple peau nous a couverts Nature:
> Pource, il faut de l'argent à couvrir nostre corps,
> Qui de luymesme est tendre & douillet par dehors,
> Auquel le chaut, le froid, & le vent est contraire.[8]

> (Nature, treating him differently from other beasts,
> Does not provide man with head-to-foot clothes.
> We see horses, lions, bears, sheep, and bulls,
> Dogs, cats, wild boars, and deer all dressed in solid skins,
> Which defend their bodies against both heat and cold.
> But Nature has covered us only with a basic skin,
> Such that to cover our bodies we require money.
> Our bodies are otherwise tender and soft on the outside,
> Too easily attacked by heat, cold, or wind.)

Nature, according to Ronsard, has not provided for humans as she/it has for nonhuman animals, leaving the former quite literally out in the cold. To compensate for their nudity, insufficiently warm skin, and generalized exposure to the elements, argues Ronsard, human animals need money ("il faut de l'argent"), that is, gold coins. The extraction of metal *ex Terra* appears, from these early moments in the poem, to be part of a system and to be required by Nature herself. In other words, Ronsard naturalizes—or Naturalizes—the exterranean by locating in Nature a lack. With this undergirding his project, Ronsard offers up a mythological vision that captures Terre's consent to the exterranean, which contrasts with the absence of voluntary provision of food and clothes for human survival. About halfway through the text, we thus come across the following vision, in which Terre opens herself up to show off to her fellow Olympians the riches she contains within:

> la Terre leur mere . . .
> Ouvrit son large sein, & au travers des fentes
> De sa peau, leur monstra les mines d'or luisantes,

Qui rayonnent ainsi que l'esclair du Soleil
Quand il luist au midy, lors que son ardent œil
N'est point environné de l'espais d'un nüage,
Ou comme l'on voit luire au soir le beau visage
De Vesper la Cyprine, allumant les beaux crins
De son chef bien lavé dedans les flotz marins.[9]

     (The Earth their mother . . .
Opened her wide breast, and through the holes
In her skin, she shows them brilliant mines of gold,
Which shine like the brightness of the Sun
When it dazzles at noon, when its burning eye
Is not hidden by the thickness of a cloud,
Or like one sees light up in the evening the fine
Visage of Venus, which lights up the beautiful hair
Of her head finely washed in the sea's waves.)

Thus we have Terre not standing in a courtroom pointing at the holes in her dress as wounds of exterranean violence, as in the *Judicium Jovis*, but proudly opening up her own body in order to reveal, "through the holes / In her skin" ("au travers des fentes / De sa peau") the golden riches within. It is no longer the miner's pick that creates the holes but Terre herself. The ex of the exterranean is here quite different, signaling a rupture and a departing of matter that Terre wants and wills. It is worth noting at this point that while much of Ronsard's poem is woven from various literary sources (especially from fragments dispersed within Stobaeus's *Florilegium*), these particular verses just quoted are generally thought to be of Ronsard's own invention, which is perhaps one reason that commentators have demonstrated a certain amount of uncertainty over how to read them.[10] The episode is sometimes—in fact usually—written off as being a mere interlude. Here, however, I shall argue that it grounds the whole text, for the way that it gives an explicitly exterranean origin to gold.[11]

A first point to note regarding these verses is that as Terre opens herself up, what she is said to show is not just *gold* but "brilliant *mines* of gold" ("les *mines* d'or luisantes").[12] Deep within Terre is matter *for humans to extract*, matter seemingly already housed in those extractive systems (mines) created *by humans*. The specificity and consequences of this word become clearer when we remember that Ronsard might have described things differently—we know that he read widely, including authors as diverse as Gerolamo Cardano, Hippocrates, and even the mining expert Georgius Agricola, all of whom left their mark on his poetic imagination.[13] Ronsard could, for example, have echoed more directly Ulrich Rülein von Calw's *Eyn wohlgeordnet und nützlich büchlein*, *wie man bergwerk suchen und finden soll* (c. 1500), known in English as *A well-ordered and useful little book about how to seek and find mines*, which describes how "mineral- or ore-creating Power" produces precious metals such as gold in "natural vessels" within the Earth known as "veins."[14] The terminology of vessels (*vasa*) and veins (*venae*) is also ubiquitous throughout Agricola's *De re metallica* (1556) to denote those places where Earth (Terra) stores minerals and metals "within her own deep receptacles [*receptacula*]."[15] Or Ronsard could have taken up phrasing employed by both mining experts and early modern alchemists about how gold "grows" in "womblike matrices" within the Earth.[16] Biringuccio, for one, speaks in *De la pirotechnia* (1540) about hills and mountains that contain "the matrices of all the most prized riches" ("le matri di tutte le piu stimate richezze") and that are the repository "of all treasures" ("di tutti i thesori").[17] The point is that Ronsard *did not* have Terre, as she opens herself up, talk of "veins" or "vessels" or "womblike matrices" but rather of "mines," that is, not of subterranean matter in those containing structures described by both alchemists and mining engineers but of matter already lodged within an extractive exterranean system seemingly just waiting to be utilized. As Naomi Klein phrased it in quite a different context, here we see Terre being "turned from mother into the motherlode."[18]

The gold that Terre reveals to her fellow Olympians is not only kept in mines, moreover, but is kept in mines that signal the matter as particularly vibrant, in the sense that Jane Bennett gives to that term.[19] The gold doesn't just *sit there*—the mines are glistening and gleaming (*luisantes*); they shine (*rayonnent*) like the bright midday sun shines (*il luit*) when its burning eye (*ardent œil*) is hidden by no cloud; they shine like Venus's face shines (*l'on voit luire*). The insistence on gold's shininess—words derived from *luire* (to gleam) appear three times in just nine verses—betoken the "metallic vitality" proper to the polycrystalline structure of metal.[20] Ronsard's verses, of course, everywhere overflow with such exuberance: shiny objects, surfaces, etc. One specific consequence of this emphasis on the vibrancy of Terre's gold mines is that—and Ronsard knew his Ovid—the association between mining and the iron age created in Book 1 of the *Metamorphoses* is here overturned. Although Terre herself points to these mines, the poet disrupts Ovidian chronology and announces: "The golden age still lives on today" ("l'âge d'or regne encore aujourdhuy").[21] The "Hymne de l'or" recognizes Terre as a generative force—as did the Gaia of the Homeric and Orphic hymns and the Terra of the *Precatio Terrae*, discussed in Chapter 1—but here her generative powers produce not plants but gold. Just as Ovid's golden-age Earth gave "without compulsion . . . all things needful," so does Ronsard's Terre—but what it gives is different.[22] As seen above, it precisely does *not* give "things needful"—if we understand by that expression things like crops and clothes—but rather gives metal that can be used to purchase what Terre/Nature does not give. Although Ronsard did not use the language of "matrices" to describe Terre's insides, it is via this substitution of metals for plants that Ronsard *does* brush up against alchemical beliefs. Bernardino Telesio in his *De rerum natura juxta propria principia* (1565) affirms that "vegetative powers" are given "to gold, silver, and the other metals" just as to plants—which is arguably what Ronsard, too, does here.[23] That Ronsard aligns Terre's fertility with gold rather than with the crops of Ceres and Bacchus is made explicit by the fact that, just before Terre shows off her gold mines, the other Olympians show off their specific treasures and, along with Zeus brandishing his lightning bolt, Mars his warrior's spear, Saturn his great scythe, Neptune his waters, and Hercules his club, the poem describes Ceres showing "her fields" ("ses campagnes"), Bacchus "his fine vineyards" ("son beau vignoble"), and Flora "her beautiful flowers" ("ses belles fleurs").[24]

The vibrancy the poem recognizes in subterranean gold obtains also within gold once it has become detached and fully exterranean. The dazzling shininess Terre points to when she opens herself up infuses exterranean gold with very real agency. Jumping around the poem, we learn that gold can cause our knees to bend, as when we kneel down before great lords; it can make objects move, putting crowns on the heads of kings, simultaneously conferring power to the ruler and causing him to act; it can induce a bookseller to part with his books to those who want to read them; it permits the sick man to call a doctor—and it allows the poet to compose poetry.[25] Somewhat like the electricity that flows through an electrical grid made up of human and nonhuman agents, gold is here confirmed to be part of a complex network of various kinds of agents.[26] The lack of it can mean nonmarriage and chastity, a literal interruption of physical and social networks.[27] Gold here is said to constitute "the blood, nerves, strength, and life of humans" ("Le sang, les nerfs, la force, & la vie des hommes")[28]—and this to such a point that "[he] who has acquired no gold in his house / Resembles a dead man [*un mort*] who wanders among the living [*entre les hommes vifz*]."[29] Gold is further said to be more enlivening than bread, wine, or fire;[30] like food, it causes bodily transformation;[31] and, in the section where Ronsard notes that gold can pay for a doctor's visit, what he first writes is—quickening the metal's efficacy—that gold "carefully looks after our body [*de nostre corps soigneusement a cure*]."[32] Within the context of this study, what is important is the continuity of vibrant metallic agency between the gold still attached to Terre and the gold once it has been mined. In other words, Ronsard's hymn ascribes the agency of gold to its being exterranean.

A final point to be made about Terre's revealing of her gold mines requires paying attention to Ronsard's Jupiter, who—unlike Niavis's, and much closer to Homer's Zeus—is depicted as reacting to the precious metals hidden within Terre not only with glee but with an unbridled eagerness to consume. The exterranean production of gold is here described as Terre's *giving* to her children, the Pagan gods, not as a *taking*. The latter ask their mother to "give them some of that shiny stuff" ("leur donner un peu de cela radieux"). Terre consents, "and, generous, honored / Her children with her gold, turning their heavens golden" ("& prodigue honora / De son OR ses enfans, & leurs cieux en dora").[33] Of all the gods described, it is Jupiter who first demonstrates a distinct desire for the exterranean product: "Thus Jupiter used it to golden his throne, / His scepter, and his crown" ("Adoncques, Jupiter en feit jaunir son trosne, / Son Sceptre, sa couronne").[34] After Jupiter, Juno creates her own golden throne, the Sun decorates her hair with gold, and then Mercury—who had been Terra's lawyer in the *Judicium Jovis*—upgrades from a wooden caduceus to a golden one: "Mercury decorated his wand with [gold], which / Before had only been of yew" ("Mercure en feit orner sa verge, qui n'estoit / Auparavant que d'if").[35] All the Pagan gods, Terre's offspring, enjoy and deploy the gold she has given, and it is Jupiter who acts first. This is not the only time in the poem, moreover, that Jupiter is described as one of gold's biggest admirers. To justify his writing the poem, Ronsard says that he will sing of this noble metal "into which even Jupiter / Metamorphoses" ("en qui mesme se change / Jupiter"),[36] alluding to the episode in which Jupiter transforms himself into golden rain in order to impregnate Danaë, an episode known from many sources, including the *Metamorphoses*, and which Ronsard himself appropriates in one of his better-known sonnets in the *Amours* to describe his desire for Cassandre: "How I wish I could, richly and turning yellow / Fall down drop by drop as golden rain / Upon the fine bosom of my beautiful Cassandre" ("Je vouldroy bien richement jaunissant / En pluye d'or goute à goute descendre / Dans le beau sein de ma belle Cassandre").[37] Ronsard also associates richness more generally with Jupiter by associating the former with the cornucopia of Amalthea, Jupiter's foster-mother.[38]

The present reading of Ronsard's vision of Terre is at odds with that of Rebecca Zorach, which sees the exterranean genealogy proposed by Ronsard as negative, even if ambiguous: "The birth of gold from the womb of Terra . . . gives it *a suspect pedigree*, embodying positive but also negative connotations of earth and earthiness."[39] Zorach does not explain why gold's being exterranean—connoted by "earth and earthiness"—would necessarily correlate with a negative evaluation. On the contrary, the "Hymne de l'or" arguably transfers (positive) vegetative power from plants to gold and has Jupiter and other gods react favorably. Moreover, this is relatively consistent with Ronsard's œuvre in general: Terre, Earth, earth, and earthiness are not, as far as this reader can tell, generally negative at all. It is precisely into this positive value of the exterranean that Ronsard taps when he brings gold within its remit. In one of Ronsard's earliest poems, a wedding song ("Epithalame") written for the marriage of Antoine de Bourbon and Jeanne de Navarre, which took place on October 20, 1548, the poet calls upon nymphs to decorate the sacred wedding bed with the richest and most colorful flowers "With which the Earth itself is painted [*Dont la Terre soit painte*]."[40] The same generous Earth is present in another early poem written on the occasion of Henri II's royal entry into Paris: As the king enters into France's capital, the poet writes, it seems as though "the Earth / Releases all of its treasures from its womb [*de son ventre*]."[41] The celebrated "Hymne de France"—the poem that first announced his major epic, the *Franciade*—invokes a similarly generous planet: In France, not only do the twin sisters Justice and Equity bloom "like two lilies," but in celebration "the Earth opens up [*la terre s'ouvre*], / And shows off its/her treasures to the new sky."[42] Further on in that same nationalistic text, Ronsard cries out: "I salute you, oh plenteous earth [*ô terre plantureuse*]."[43] Here, as in the "Hymn to Gold," the Earth is fertile and generous, offering up its riches—flowers in the "Epithalame," unnamed treasures in the "Hymne de France," and gold in the "Hymn to Gold." Many other examples of similar usages and such positively connoted references to Earth, earth, and earthiness could be adduced. Within the confines of the *Hymnes*, of the 102 times the word *Terre* is used, only rarely does it connote anything negative.[44] In the "Hymn to Gold," Ronsard switches these various floral and agricultural bounties for gold.

SITES OF EXTRACTION

Switching from whole to part shifts our attention from this vision of Terre opening herself up and showing brilliant mines of gold within to not only the ways in which Ronsard fabricates a mythological exterranean origin for gold but also how he pictures specific sites of extraction. At one point, Ronsard imagines a detractor who mocks him for his praise of gold by alluding to its potentially humble exterranean origin within the sands carried by streams:

> Quelcun apres cecy me viendra dire encor
> Comme par moquerie: hé mais qu'esse que l'OR,
> Pour en faire un tel cas, qu'un sablon que l'on treuve
> Aux rives de la mer, ou sur le bord d'un fleuve?
> Il ne chait pas du Ciel, il faut avec grand soing,
> A qui le veut avoir, l'aller chercher bien loing.[45]

> (Someone after this is bound to come say to me,
> As if in mockery: so what is this GOLD

About which you make a fuss, if not sand found
On the seashore, or else on the bank of a stream?
It does not fall from the Sky. He who wants it
Must take great pains and search for it far away.)

We might not take this mention of river gold too seriously, that is, too literally. After all, and as Paul Laumonier recognized many years ago, Ronsard
here *merely* imitates a fragment from the Greek gnomic poet Naumachius conserved in Stobaeus's *Florilegium*,[46] a fragment Ronsard likely found in Conrad
Gesner's edition, *Sententiae ex thesauris Graecorum delectae* (1543), where it is
given (in Gesner's Latin translation) as:

Aurum & argentum nihil quàm puluis sunt, & qui reperiuntur,
Lapilli in calculosa crepidine maris.
Multi autem iacent ad ripas fluuiorum temere.

(Gold and silver are nothing more than dust, like the stones that can be
found on the pebbly beach, or else on the banks of rivers.)[47]

There are several reasons to pause over these verses and read them more materially than has been allowed. To begin, it has not—to my knowledge—been
noted that these very same lines from Naumachius were also quoted by Georgius Agricola in his *De re metallica*.[48] Whether or not Ronsard knew this, he puts
himself in the same position as the mining expert. As we shall see in Chapter 3,
Agricola frequently adduces citations that he hears being used by those who are
against mining, in order to refute them. Ronsard, here, does exactly the same,
for he quotes these lines as if someone (*Quelcun*) were addressing him in a mock-
ing tone, to question the *bien-fondé* of his praise of gold. Beyond the fact that
Ronsard and Agricola quote the same lines from Naumachius, the interesting
aspects of this mention of river gold are, first, that this was the *major* source of
gold up to the sixteenth century—until the massive-scale importing of mined
gold from the New World, which Ronsard references here only obliquely via
the mention of gold coming from far way (*bien loing*)—and, second, that the
poem itself specifically foregrounds, in ways we shall see, river gold.

Agricola gives over a huge section of Book 8 of *De re metallica* to talk about
placer mining, that is, the extraction of gold from river sands using methods
such as panning or the use of a rocker or sluice box (Figure 7). And while Ronsard seemingly rejects the idea, expressed in the fragment from Naumachius,
that gold might be compared to "sand found . . . on the bank of a stream," in his
Bergbüchlein (c. 1500), Ulrich Rülein von Calw states that gold "occurs in different ways," including "in ordinary river sand" ("in schlechte sandt des flyeß")[49]
and that that particular source of gold produces, he claims, "the purest and

Aſſer A. Tabellæ B. Filaferrea C. Anſæ D.

FIGURE 7. Placer mining in Agricola, *De re metallica*. With kind permission of the Rare Book and Manuscripts Library, Columbia University.

most exalted kind [*das aller klarist vnd hochste gold*] because its matter is most thoroughly refined by the flow and counter-flow of the water" and because of "the characteristics of the location where such gold is found, that is, the orientation of the river in which such placer gold is made."[50]

The "Hymne de l'or" takes river gold seriously in an even more pressing way, however, adducing a local site of extraction. The latter appears only very briefly, at the start of the poem, but arguably haunts the whole text in the way it articulates a material connection between the poem's verses, its dedicatee, shiny gold, and a particular, localized, artisanal site where that metal can be found, a site, moreover, enmeshed in the history of French Renaissance art. It is worth quoting the verses in question in full.

**HYMNE
DE L'OR,
A JEAN DORAT.
VERS HEROIQUES**

Je ferois un grant tort à mes vers & à moy,
Si en parlant de l'OR, je ne parlois de toy
Qui a le nom doré, mon DORAT, car cet Hymne,
De qui les vers sont d'or, d'un autre homme n'est digne
Que de toy, dont le nom, la Muse, & le parler
Semblent l'or que ton fleuve Aurence fait couler.[51]

**HYMN
TO GOLD
FOR JEAN DORAT
HEROIC VERSES**

I would greatly wrong my verses and myself
If, when I speak of gold, I spoke not of you
Whose name is golden, dear Dorat, for this Hymn,
Whose verses are made of gold, is worthy only
Of you—your name, your Muse, and your speech
Resemble that gold carried along by the Aurence.

The "Hymne de l'or" thus opens with multiple gold translations, both linguistic and material, which give a more positive view of gold extracted from river sands. Ronsard dedicates his poem, from its very title, to Jean Dorat (1508–1588), his former teacher at the Collège de Coqueret. The first six alexandrines then restate that dedication in even more glowing terms, making explicit the onomastic play on words—Dorat has "a golden name" ("le nom doré"). Discussion of this dedication of a poem about gold (*or*) to a poet with a golden name (*doré/Dorat*) generally focuses on biographical connections: Both Ronsard and Dorat were *quémandeurs*, seekers of sponsorship, but Ronsard was arguably better at winning royal gold—he thus perhaps wanted to teach his teacher.[52] But I think there is something much more interesting happening here. Purely biographical readings eclipse the material: here, Ronsard's direct mention of the river Aurence, a tributary of the Vienne just upstream from the city of Limoges, where Dorat was born in 1508 and where he lived until leaving for Paris at age nineteen. In the sixteenth century, gold-washing sites (*laveries d'or*) would have been dotted along the river for several miles, and sights like those depicted in Agricola (Figure 7) would have been quite common.[53]

In these opening verses, Ronsard weaves together the "Hymn" we are reading ("whose verses are made of gold") with his dedicatee, whose name, Muse, and speech resemble the gold "carried by the Aurence." Ronsard is, in part, drawing on an association also known to others: According to one of his contemporaries, Jean changed his name from Dine-matin to Auratus ("golden" in Latin) and thus to Dorat (in French) precisely because he learned to write poetry on the banks of the Aurence.[54] Anyone familiar with Pléiade poetry would be forgiven for suspending analysis here: Ronsard's (and Dorat's) verses are nothing if not richly adorned and overflowing with ornament. And indeed it was Dorat who taught Ronsard how to "fain and hide fables" underneath poetry's "fabulous cloak."[55] Ronsard was, moreover, not the only of his contemporaries to link Dorat, the Aurence, and gold: François de Belleforest, when he describes Limoges's wateriness, also addresses praise to Dorat: "Of all the excellent men from Limoges, I cannot not speak, nor omit, this Gallic Homer and Greco-Latin Pindar, Jean Dorat, the most rare and subtle Poetic spirit of our century."[56] What I want to emphasize is that whatever the biographical *cause* of this association between Dorat, *or*, and the Aurence, important here is that the poem opens not with the planetary vision of gold-rich Terre but with this (albeit) brief mention of an auriferous river where early modern professional and amateur *orpailleurs* would look for gold.[57]

The Aurence looms large in early modern descriptions of Limoges. François de Belleforest says the city was "very well watered, and indeed aquatic"; it is, he adds, "abundantly watered," such that the horses do not go thirsty and the streets are always clean.[58] Indeed, the river winds around the city, making it somewhat omnipresent. Coming into existence at the foot of the Mounts of Ambazac to the north of Limoges, it winds its way east and south, then south into Lake Uzurat, then on around Limoges, passing through the Parc du Moulin Pinard (Figure 8), the Parks of the River Aurence I and II, the Park of Mas Jambost, and on until it finally flows into the Vienne at Aixe-sur-Vienne. The river still organizes how walkers, joggers, picnicking families, and "nature" lovers inhabit the town. Its status as potentially gold bearing has seemingly been forgotten, however; it is not mentioned in contemporary guidebooks, and anglers are many times more numerous than *orpailleurs*. That said, it is not difficult to see why Dorat and others appreciated the river's special relationship to gold—even today, the river bears gold specks and gold-speckled stones. Whether actual gold or only pyrite (FeS_2), the river certainly overflows with shiny gold-colored matter (Figure 9).

Ronsard's mention of the Aurence opens out not only onto the gold-bearing river and onto Dorat, who wrote verses sat next to it, but also onto a particular artistic product of the town: Limoges enamels. As the author of the first national French atlas, Maurice Bouguereau, wrote, the town "abounds not only in good artisans of all types" but also especially in craftsmen gifted in

FIGURE 8. The river Aurence in the Parc du Moulin Pinard, Limoges (2015). © Phillip John Usher.

FIGURE 9. Gold-colored specks and rocks fished from the river Aurence at the tip of Lake Uzerat in July 2015. © Phillip John Usher.

"painting done in enamels [*peinture faicte en esmail*]."[59] Enamel, not gold, was of course the main ingredient here, but gold was also used, extensively, in Limoges enamels, especially to accentuate key elements of the images, typically haloes, hair, clothing outlines, stars, vegetation, animals, architectural features, or other (usually small but important) details.[60] Gold from the Aurence might also have been used in the production of reliquaries, candlesticks, croziers, buckles, brooches, belt buckles, and a whole host of other golden objects.[61]

WHOLE AND PART

On the one hand, then, we have Terre, who opens herself up to show mines of gold. We might call that vision global or say that it is imbued with a "sense of planet."[62] That Terre is a reprisal of Gaia, but refashioned, a Terre that is not the Terra of Ovid's iron age ripped apart by miners but a new version of golden-age Terra who provides not just crops but gold. On the other hand, we have this local site of artisanal extraction, the river Aurence—the local, offering a "sense of place." Or we might phrase this differently, saying that, by juxtaposing these two visions, the poem invites the reader to think of the two visions as inherently interlinked and of gold as being part of a global system (Terre) yet made available at a local site (the Aurence), which brings us into the domain of what Robert Robertson, to grasp at this collapsing of the antinomy between, or "reflexive synthesis" of, the global and the local, calls the *glocal*.[63] For Robertson, the local and the global are in constant dialectical exchange—they are complicit, as in certain contemporary products produced by global chains (for example, McDonald's) but adapted for local audiences, such as the McDonald's *McWrap Chèvre*, meant as the fast-food version of France's traditional *salade de chèvre chaud*. This is, perhaps, what is happening here: The river Aurence localizes Terra's monstration of her massive and "brilliant mines of gold," *and* Terra globalizes the Aurence's "carrying along" of small *paillettes* of gold. The *and* is crucial here: Despite the radical difference between an actual French river and Terre, there is arguably no tension between the two, for the poem posits a dialecticity that advances a material genealogy. The gold in the Aurence is a glocal phenomenon, in that the reader can presume this local gold to have also been produced by *Terre*, in other words, that in this particular place in France, Terra-produced gold has been indigenized—it is no longer used by Olympians but by Limoges artisans.

The point where such thinking breaks down, however, is the very mysterious mention of gold that has to be sought out "far away": "[Gold] does not fall from the Sky. He who wants it / Must take great pains and search for it far away" ("Il ne chait pas du Ciel, il faut avec grand soing, / A qui le veut avoir, l'aller chercher bien loing").[64] As we will see in much more detail in the following chapter, the production and availability of gold change drastically in the sixteenth century following the discovery and exploitation of gold mines in the Americas. As an indication of this change, we can note that during the century, approximately 150 tons of gold and 7,500 tons of silver would be shipped from the New World to Europe.[65] The "Hymne de l'or" offers up a mythological vision to capture the exterranean origin both of gold and of gold's vibrancy, and it also localizes gold production in the river Aurence that runs through Limoges—but Ronsard alludes only very briefly to gold that is mined with effort, far away, as if specifically *not* to mention the fact that gold is being taken from the New World by workers operating under nearly slavelike conditions. This silencing becomes even more striking given that elsewhere in this same poem, Ronsard specifically mentions *other* New World imports and compares them to gold:

Qu'on ne me vante donc ce gayac estranger
Par dessus ce metal qui sauve du danger
Roys, Princes, & Seigneurs, soit que bouilly le boyvent,

Soit qu'autrement par luy douce santé reçoivent.
Il n'est pas seulement de nostre corps soigneux,
Il l'est de nostre esprit, qui, tant soit chagrigneux,
Despit, triste, pensif, resveur, melancolique,
Est tout soudain gary d'une douce musique,
Ou de livres nouveaux divinement escris
Que l'OR nous donne, à fin d'alleger nos espritz.[66]

(May no one praise to me that foreign guaiacum plant
Over this metal, which preserves from danger
Kings, princes, and lords—they boil it and drink it
Or in other ways receive from it sweet health.
It cares not only for our body, but
Also for our mind which, however sorrowful,
Hurt, sad, pensive, day-dreaming, or melancholic,
Is suddenly cured by sweet music,
Or by new books divinely written,
Which GOLD gives us, so as to lighten our spirits.)

Ronsard refers to the guaiacum plant labeled specifically as "foreign" (*ce gayac estranger*), a flowering plant native to subtropical and tropical regions of the Americas that explorers had recently brought back to Europe as part of the wider Columbian exchange. If Ronsard wishes specifically *not* to hear the plant's praises sung, it is because many voices were indeed doing just that, somewhat generally as a panacea and more specifically as a cure for syphilis, as in the *De guaiaci medicina et morbo gallico* (1519), the work of the German humanist Ulrich van Hutten, who claimed both to have suffered from the disease and to have been cured by guaiacum.[67] Luca Ghini's *Morbi gallici curandi ratio perbrevis* (written 1548–1555) and many other texts of the period are equally adulatory.[68] Several decades before Ronsard composed his "Hymne de l'or," guaiacum and other American woods thought to have curative properties were being imported to Europe in large quantities. Ronsard's term *estranger* is apposite: "Whether or not the New World was believed to be the origin [of syphilis], it was undoubtedly the source of a cure."[69] Ronsard's dialectical material genealogy—Terre/Aurence—in which gold is always glocal thus also serves to draw attention away from, even to obfuscate, the very real material origins of early modern gold in the Americas.

CHAPTER 3

TERRA GLOBALIZED

The figures of Terra examined in Niavis's *Judicium Jovis* and in Ronsard's "Hymne de l'or" ultimately belong much more to mythology than to geography. Even when interrogated by or juxtaposed next to specific sites of exterranean production—the Erzgebirge, the river Aurence, etc.—Terra remains somewhat spatially aloof. Contemporary to these figures of Terra are the "discovery" of the New World in 1492, the production of the first terrestrial globes, the rediscovery, translation, and wide print dissemination of Ptolemy's *Geographia*, as well as the wider growth of humanist geography at local, national, and planetary scales, all of which contributed to what might be called a globalization or a planetarization of Terra.[1] As of 1507 it was possible to purchase Martin Waldseemüler's cutout terrestrial globe gores and to hold Terra in one's hands and contemplate it—as we've already heard Latour put it—"like a god" (Figure 10). Similarly, in his mid-sixteenth-century *Cosmographia*, Sebastian Münster would write that all knowledgeable persons are aware that "terram [est] globum magnum & rotundum" ("Terra is a big and round globe"), as illustrated in several woodcuts (Figure 11).[2] During this period, the mythological Terra and the geographical-globalized Terra clearly coexist. The present chapter turns to this second figure of Terra to see what sense of the exterranean emerges. One way to think of what follows is as an attempt to replanetize Ronsard's vague mention that extraction happens "bien loing" (in some far off place).

The turn from mythological to geographical-globalized Terra here coincides with, and is of particular interest because of, another more recent sense of disconnection. The Anthropocene is frequently criticized for the fact that its sense of the "whole"—that is, its implication of humanity/a species/the *anthropos*—is blind to everything different across the planet's billions of occupants.[3] Through its lens, the CEO of a major oil company and a climate refugee fleeing Bangladesh

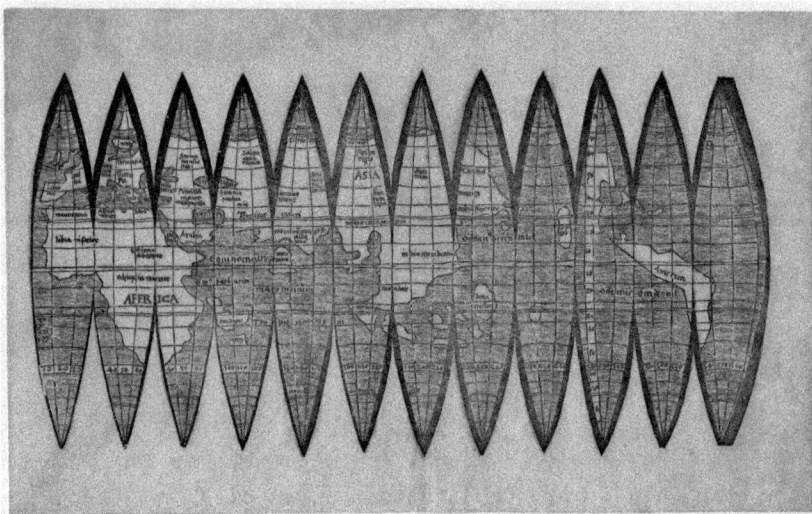

FIGURE 10. Martin Waldseemüller's terrestrial globe gores (1507). James Ford Bell Library, University of Minnesota.

look identical. The problem, of course, is not academic. Each day, as the newspapers inform us of events wherein weather meets climate, we read sentences such as this one: "Most victims" of the 98-degree-Fahrenheit heat wave in Karachi, Pakistan, were "elderly or poor."[4] In a nutshell: *We* may be a geological force, but *we* are not all equally responsible for that fact, nor are *we* all likely to experience in the same way or to the same extent the negative consequences thereof. The figure of the *anthropos* and the human bodies that populate the planet are disconnected, a realization central to the climate justice movement, to Pope Francis's discussion of our "common home" in his 2015 *Laudato Si'* encyclical, and to the work of certain postcolonial scholars such as Dipesh Chakrabarty.[5] In the pages that follow, I thus hope to trace out some of the ways in which the difficult connection between two figures of Terra collides with the Anthropocene's own struggle to talk about both the *anthropos* and real human individuals. One key point is that adding geography back to Terra means seeing how populations are always situated.

The European "discovery" and colonization of the Americas was motivated by the hope for and resulted in the large-scale extraction of raw materials, especially gold and silver, which were shipped to Europe in ever greater quantities. There is thus a direct historical and causal connection between the reshaping of planetary geography in the early modern period and extraction, that is, between the changing face of Terra and new senses of the exterranean.

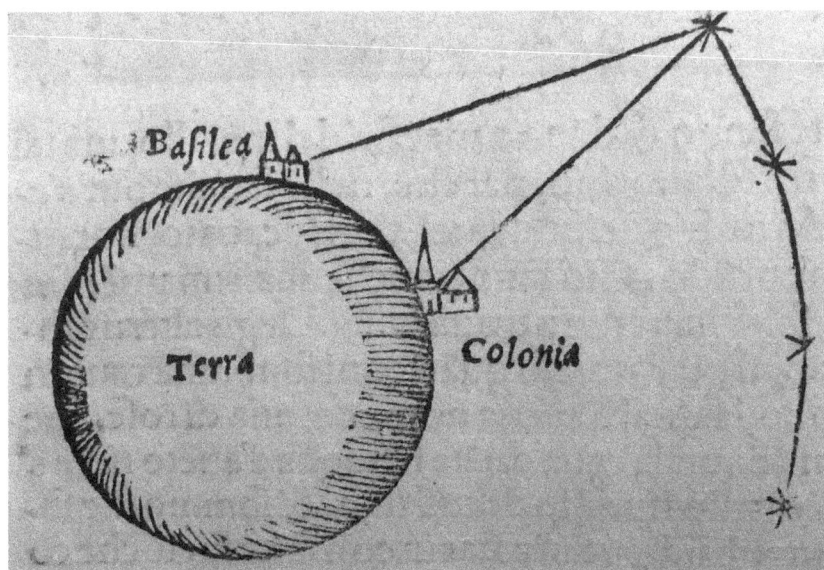

FIGURE II. Terra as globe in Münster, *Cosmographia*. *Cosmographia vniuersale* (Colonia: appresso gli heredi d'Arnoldo Byrckmanno, MDLXXV [1575]). With kind permission of the Rare Book and Manuscripts Library, Columbia University.

As we shall see, this newer sense of the exterranean—related to the *bien loing*—necessarily also implies giving the human agent of extraction a particular, foreign, face. To grasp this interconnection of geographical-global Terra and the exterranean in this period, it is useful to open with fresh eyes the *Cosmographia* of Sebastian Münster, an atlas of double-sheet maps followed by six long books of text to which 120 informants contributed. It has been called "one of the great publishing successes of the sixteenth century" precisely because "for the first time" it allowed a wide swath of educated readers to *see* Terra.[6]

There can be little doubt that the *Cosmographia* played a major role in developing an image of Terra as a big globe more or less coextensive with the ecumene (Gk. οἰκουμένη), or known/inhabited part of the Earth. The huge volume gives a fairly comprehensive idea of Terra's size and shape following the discovery of the New World in 1492. Chapter 16—"De magnitudine terrae"—takes up the topic of Terra's size, stating how knowledgeable persons know "terram esse globum magnum & rotundum."[7] The following chapter takes up and redeploys the most common, tripartite, medieval description of Terra, focusing on the "primary division of the whole of Terra into Europe, Africa, and Asia" ("prima divisione totius terrae in Europam, Africam, & Asiam").[8] Other parts of the *Cosmographia* graft the New World to this former structure,

FIGURE 12. World map in Münster, *Cosmographia* (1575), n.p. With kind permission of the Rare Book and Manuscripts Library, Columbia University.

such that the Americas are integrated into this conception of the whole Terra (*tota Terra*) via the world map included at the start ("Typus Orbis Universalis," or, in the Italian version shown here, "Figura del mondo universale") (Figure 12); the map of the New World ("Tabula novarum insularum"), also included in the initial collection of maps; as well as the significant number of pages in Book 5, gathered under the title "On the New Islands, and on How, When, and by Whom They Were Discovered" ("De novis insulis, quomodo, quando, & per quem illae inventae sint"), complete with its descriptions of Columbus, cannibalism, and other now familiar topics.[9]

Münster's *Cosmographia* has, on the one hand, received a lot of attention for its role in producing our post-1492 idea of Terra. On the other hand, few (if any) readers have paid much attention to *how* the work begins. Before the reader gets to Chapter 16 about Terra's size, or to Chapter 17 about the three continents into which Terra was traditionally divided, or to the pages about the New World, she first reads a number of chapters that directly describe not Terra but the exterranean: Chapter 8 ("De metallis, potissimum de auro") discusses metal and especially gold; Chapter 9 ("De argento") focuses on silver; Chapter 10 ("De argento vivo") discusses mercury, essential to mining; Chapter 11 ("De aere") turns to bronze, Chapter 12 ("De ferro") to iron, and Chapter 13 ("De mixturis metallorum") to alloys. Next, Münster dedicates Chapter 14 to the topic of where metals are extracted in Europe ("Quibus in locis Europae olim fuerit effossum, & hodie foidatur metallum") and Chapter 15 to the topic of the machinery and human agency involved in extraction ("De machinis, quibus metallarii utuntur in fodinis").[10] Thus, in this work acknowledged to play a major role in defining the post-1492 sense of Terra, sandwiched between a post-Columbian map of the planet, continental maps, including one of the Americas, and chapters about the size and shape of Terra are these chapters about various aspects of exterranean activity. Terra may be a "globe," but it is clearly, here, a globe defined by and of interest because of its surface depths and the exterranean activities to which the latter are essential. The prominence of extraction at the beginning of the *Cosmographia* is made all the more present by woodcuts such as the one reproduced in Figure 13, which shows various stages of the process.[11]

LOOKING FOR INDIGENOUS MINES

To get a fuller grasp of how whole (Terra) and part (exterranean activity) fit together in the context of this globalized Terra, with an eye solidly fixed on the more recent challenge of the unachoredness of the *anthropos*, let us turn to one of the most canonical texts of early modern France, Michel de Montaigne's chapter "On Coaches" ("Des coches"), published in the third volume of his *Essais* (1588). Frequently discussed in terms of its excoriating critique of colonialism, the chapter also—and this is noticed much less often—emphasizes how exterranean fantasies drive the colonial impulse: "While sailing along the coasts *on the lookout for the natives' mines* there were some Spaniards who went ashore in a fertile, pleasant, and densely populated countryside" ("En costoyant la mer *à la queste de leurs mines*, aucuns Espagnols prindrent terre en une contrée fertile et plaisante, fort habitée").[12] Europeans in the New World are looking primarily, according to Montaigne, for already existing sites of extraction—and what they find, at least at first, is fertile countryside. All sorts of Edenic Falls are latent here, of course. This sentence in "Des coches" has generally been overshadowed by what follows: the virulent, biting, and indeed epoch-making indictment of colonial cruelty in the Americas. Here, I want to underscore the fact that that cruelty has its origins, within the chapter, in this longing for mines. More specifically, when Montaigne writes that Europeans seek out "natives' mines" ("leurs mines"), he refers to the mines belonging to "so many millions of individuals" ("tant de millions de peuples") who would be "put to the sword" ("passez au fil de l'espée").[13] The massacre of New World

FIGURE 13. Mining in Münster, *Cosmographia* (1575), 7. With kind permission of the Rare Book and Manuscripts Library, Columbia University.

populations is, from the very beginning in Montaigne, shackled to a dream of extraction in which the body of the Earth and the bodies of New World populations become intensely commingled.

Creating a bookend for this initial moment, Montaigne replays this search for indigenous mines of gold in the closing section of the chapter, where the final victory of the Spanish over indigenous populations is depicted via the (literal) fall of the last king of Peru, Atahualpa (sometimes spelled Attabalipa)

(c. 1500–1533), whom Cortés and his men captured at the Battle of Cajamarca in 1532. Montaigne's description of that moment, marked—as Richard Sayce put it—by "understatement and lack of overt emotion,"[14] is as follows:

> Ce dernier Roy du Peru, le jour qu'il fut pris, estoit ainsi porté sur des bran-cars d'or, et assis dans une cheze d'or, au milieu de sa bataille. Autant qu'on tuoit de ces porteurs pour le faire choir à bas, car on le vouloit prendre vif, autant d'autres, et à l'envy, prenoient la place des morts, de façon qu'on ne le peut onques abbatre, quelque meurtre qu'on fit de ces gens là, jusques à ce qu'un homme de cheval l'alla saisir au corps, et l'avalla par terre.

> (The day that he was captured, that last King of Peru was in the midst of his army, borne seated on a golden chair suspended from shafts of gold. The Spaniards in their attempts to topple him [as they wanted to take him alive] killed many of his bearers, but many more vied to take the places of the dead, so that, no matter how many they slaughtered, they could not bring them down until a mounted soldier dashed in, grabbed hold of him and yanked him to the ground.)[15]

Montaigne saves describing this key moment in the history of the Americas for the very end of his chapter, reversing the order of events as they are de-scribed in his main source text, Francisco López de Gómara's *Historia general de las Indias* (1552), which he read in French translation.[16] His brief and pithy summary collapses into just a few lines the topic of transport highlighted by the chapter's title; the death of Atahualpa; the courage, resilience, and loyal dedication of Atahualpa's men; and the cruelty of the Spaniards. As Timothy Hampton has noted, the figure of the falling Atahualpa functions temporally (to signal "the end of indigenous empires in the New World") and spatially (to figure "the relationship of conqueror and conquered").[17] But Montaigne's summary description also highlights exterranean gold. The passage, in Tom Conley's words, "has a strange aura in its reflection of gold (*or*) in words that would slake the Spaniards' desire to topple the monarch from his litter."[18] In-deed, in describing the means of transport employed by Atahualpa, Montaigne twice uses the word "gold" ("or") to describe both the litter's carrying poles of "shafts of gold" ("brancars") and the "seat" ("cheze") itself. There is nothing extraordinary in this: Gómara also describes the presence of gold. Telling, however, is that Gómara also mentions other luxurious details: In the *Histo-ria general de las Indias*, Atahualpa's litter is for sure decorated with gold—but also with feathers of various colors ("[des] plumes de perroquets de diuerses couleurs," in the French translation that Montaigne read); Gómara's chair was also golden but additionally adorned by a rich wool cushion decorated with

FIGURE 14. Théodore de Bry, Atahualpa, from *Americae*, Part VI (1596). BPK Bildagentur / Kunstbibliotek / Art Resource, NY.

precious jewels ("vn riche coussin de laine garni de fort beaux, & precieux ioïaux").[19] Montaigne deletes these and other details, retaining only the goldenness of the litter, thus singling this aspect out and establishing a direct connection between Atahualpa's fall and the orginary search for mines. Montaigne's emphasis here is even clearer when we compare his paragraph to contemporary illustrations of the same scene: If we look at the woodcut printed on the cover of *La conquista del Perú* (1534), a similar one on the cover of Francisco de Xerez's *Verdadera relación de la conquista del Perú* (1534), or one by Théodore de Bry (Figure 14), we see no particular emphasis on gold. The specificity of Montaigne's text becomes clear: A longing for extracted matter dwells at the heart of the search for new parts of and construction of economic connections upon the *orbis terrarum*. As in Münster's *Cosmographia*, the new post-1492 globalized Terra implies first and foremost the exterranean.

Critics disagree as to whether Montaigne read Bartolomé de las Casas's *Brevísima relación de la destrucción de las Indias* (*Brief Account of the Destruction of the Indies*) (pub. 1552).[20] Whether or not he did, it is clear that Las Casas's *Brevi-*

sima relación equally commingles New World genocide and European fantasies of extraction. Las Casas's celebrated text adumbrates and denounces in no uncertain terms Spanish cruelty in the New World—and it does so throughout by linking Spanish cruelty to New World mining. Las Casas mentions gold and silver mining on numerous occasions, almost always by connecting the exterranean with the human labor required to perform extraction, such that the mined land and the exploited bodies come to be related both metaphorically and metonymically. For example, when describing how the shipments of high-quality gold from the New World to the Old almost filled Spain to overflowing, Las Casas writes that this gold was extracted "from the bowels of the Earth" ("de las entrañas de la tierra") by the "sweat of native [workers] who toiled down the mines and . . . perished there" ("donde, como se dijo, murieron").[21] The exterranean captures here the movement of matter *away from* the Earth and the simultaneous burial or *enterrement* of the bodies of those human agents who work to extract it. We hear an echo here of Fortuna's response that exterranean activity will also imply the miner's body being swallowed up by Terra ("corpus vero eius a terra conglutiri")[22]—but the difference is that Las Casas's human agent is particularized as being the victim of a colonial system. Similar linkages of the exterranean to the body of the indigenous miners and to the colonial project in its entirety are omnipresent throughout the *Brevísima relación*, for example in passages about mining in Cuba, the Kingdom of Yucatán, and elsewhere.[23] One image captures this interconnection with concision: At one point in a chapter about the "Kingdom of Venezuela," Las Casas writes of how the Spanish "seized the king of the whole province and set about wringing gold out of him by torture [*por saccalle oro dándole tormentos.*]"[24]

In the *Brevísima relación*, the extraction of gold from the "bowels of the Earth" ("de las entrañas de la tierra"), of labor from indigenous miners, and the devastation of landscapes are all intimately interlinked. As individual bodies are destroyed, so too are the territories and landscapes they inhabit, above and below ground. For example, Las Casas evokes in the same breath the "unfortunate territories" and "their innocent inhabitants";[25] he talks of the destruction and violence that "all these territories have witnessed and their people suffered."[26] Devastated (often mined) landscapes resonate with close-by descriptions of murdered bodies. Talking about the provinces of Popayán and Cali, which have been "ravaged and ruined," Las Casas describes the pain provoked by the sight of the newly devastated landscape: The "land [was] extremely fertile and beautiful," but now one can only be "full of the pain and sorrow" when seeing the area.[27] In some of these towns, one is faced "with the sorry spectacle of fewer than fifty people scrabbling among the remains of what was once a town of a thousand or two thousand people." Other settlements are now "totally ruined and abandoned." And then there are "areas, once teeming with people, where the only sight for a hundred leagues, or even two or three hundred, is of scorched earth and the blackened ruins of towns and cities."[28] The German merchants to whom the Spanish Crown handed Venezuela in 1528 were responsible, writes Las Casas, for "devastating over four hundred leagues of the most fertile and blessed land on earth, and for killing all the people or driving them out of great provinces that once struck awe in the beholder: valleys forty leagues in extent, whole regions as delightful as one could desire and settlements as rich in gold as they were populous."[29] The chapter concludes with a final reminder of the devastation: "The provinces of Venezuela, together with the more than four hundred leagues of territory totally devastated and ruined in the way we have seen, formed the most heavily populated area on earth as well as the most prosperous and the richest in gold."[30]

European travelers did not bring anthroturbation to the New World. And even if they indeed brought new techniques, they did not import the desire for or knowledge about gold and silver mining, as any one of the golden or silver Pre-Columbian artworks not melted down by Europeans into ingots clearly confirms—see, for example, the rich collection of such objects at the Museo oro del Perú.[31] Nevertheless, if Europeans traveled to the Americas, it was—as Montaigne and Las Casas clearly show—motivated by protogeological fantasies of *large-scale* extraction of precious minerals. And they indeed imported new processes and practices and so doing laid waste to a continent and millions of people. Both mining and coerced indigenous labor began, as one scholar writes, "almost immediately after the arrival of the Spaniards and their conquests in the early 1530s," quickly making mining "one of the most important activities in the New World."[32]

EXTERRANEAN LABOR

For a further perception of what the exterranean becomes in the Americas, when Terra is fully geographical, let us examine a woodcut that shows Cerro Rico, a mountain in Potosí (present-day Bolivia) (Figure 15).[33] First published in Seville, the heart of the European gold trade, in an edition of Pedro Cieza de León's *Parte primera de la chronica del Perv* (*Chronicles of the Conquest of Peru*) (1554),[34] this woodcut was the very first published image of the mountain and thus the very first view most Europeans would have had of the richest argentiferous mountain in the world.

Cerro Rico stands, majestic, cut into by near-surface-level veins, as if to attract European investors or adventurers. At the bottom left of the image,

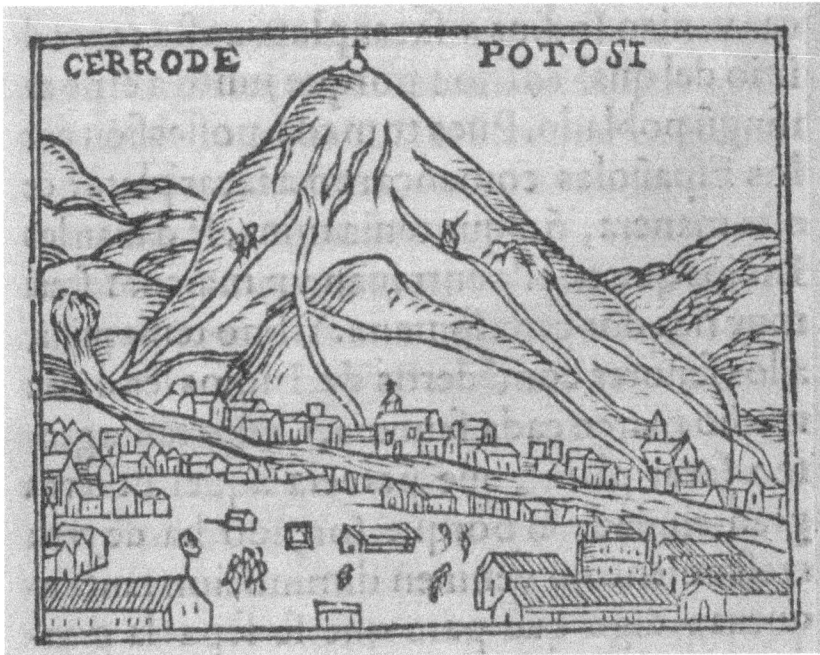

FIGURE 15. Potosí, in Pedro Ciezca de Leon, *Parte primera de la Chronica del Perv* (1554). Beinecke Rare Book and Manuscript Library, Yale University.

we see a head on a pole representing the supposed presence of Justice, that is, how the Spaniards claimed to have brought with them not just religion (see the cross atop the mountain) but also *policía* (good governance and justice).[35] The text of the *Crónica* inscribes the mountain within a Christian narrative: "We Christians ought to rejoice and give thanks to our Lord God that, in so great a country, so distant from our Spain and from all Europe, there is such justice and such good government, with churches and houses for prayer in all parts."[36] The text tells the story of the supposed "discovery" of the extraction site: "A Spaniard named Villaroel was searching for veins of metal with some Indians, when he came upon this wealth in a high hill, being the most beautiful and the best situated in all that district."[37] Mining had, of course, been taking place on Cerro Rico for a long time—but the arrival of the Spanish marked a turning point toward (1) huge-scale extraction and (2) *othered* labor. The woodcut in Figure 15 was thus the positive *ad campaign* image of New World mining, an image moreover reused many times and throughout Europe, for example in the 1581 English translation of Agustín de Zárate's *Discoverie and Conquest of the Provinces of Peru . . . And also of the ritche Mines of Potosi.*

FIGURE 16. The silver mine in Potosí, from Théodore de Bry, *America (1590)*. Snark / Art Resource, NY.

In an image included in his famous work *America sive collectiones peregri-nationium in Indiam Occidentalem*, aka *Great Voyages* (1590–1634), the Flemish Calvinist Théodore de Bry portrays Potosí's mountain quite differently (Figure 16), pulling attention away from the mountain's (potential) riches and toward the very real labor involved in extracting them.[38] The woodcut testifies to the huge scale of human terraformation, sharing much with the kinds of extraction landscapes we find in the photographs of Edward Burtynsky. As we look at De Bry's woodcut and Burtynsky's photographs, we are simultaneously drawn into and excluded from these scenes of exterranean labor.[39] We are dwarfed by the size of the mountain, but we are also strangely forced to recognize the human origin of the labor responsible for its shape. Documentary, *unheimlich*, and elegiac, these images capture precisely the turn from the *anthropos* as a geological force upon Terra to specific human groups caught up in a complex system within which and somewhat despite themselves they terraform.

Figures 15 and 16 also function differently. The main disruption to the land-scape in De Bry's woodcut (the sliced-open mountain) is represented in a way that is not fully realistic, in that the mountain is not supposed to be thought as *really* sliced open like this. Nor is the cut into the mountain—in narrative terms—diegetic, for the "men at work" are not the authors of this "slice into" the landscape. They might be hollowing out the mountain but clearly did not cut out one of its sides. In other words, whereas Burtynsky in his *Quarries* shows landscapes that are the result of anthroturbation, the major landscape intervention in De Bry's image is that of the engraver: *He* has cut the mountain open to show us the exterranean labor that takes place inside. The emphasis is here not on the *results* of human labor but on the *conditions* of that labor. With this image, we are invited to think through the complex networks that link visible human agents (the miners), invisible human agents (the conquistadors and the mine managers), and nonhuman agents (technology, the geology of this particular geographic area, etc.) into discourses that concern extraction, colonial power, and justice.

De Bry's woodcut overflows with small details. The miners whom De Bry represents are naked native laborers who, so contemporaneous accounts tell us, would normally spend three or four days at a time underground, chipping out precious nuggets deep in the bowels of Cerro Rico.[40] We see a wide ladder, made of thick wooden rungs slotted into braided rawhide, wide enough for workers to move in both directions at the same time.[41] Workers descend on the left, and on the right carriers ascend, carrying out the mined ore. Laws from 1574 stated the maximum size and dimensions of such ladders (eighty-two and one-half feet in length; rungs were sixteen and one-half inches apart).[42] As we look at the ladder, we should recall that at least one person was seriously wounded per day on the ladders, either through falling or by being struck by ore that fell from the carriers above.[43] We see miners holding tallow candles—on the left, they hold them in their left hand; on the right, in their right hand, thus foregrounding these candles as much as possible. The presence of these candles is not just verisimilar (to allow the viewer to see the miners at work)—they were at the heart of the extraction process and of miner survival. Moreover, records show that candles were generally of poor quality and supplied in insufficient quantity, meaning that the miners (whatever their status) often had to spend their own money to purchase extra candles. When we see the candles, we should thus think of the miner holding them, who would have known how much that light was eating into his already meager income.[44] Outside the mountain, to both the left and right, we see llamas being directed by raised arms, a reminder of the circulation of supplies and extracted materials, of the economic and market as well as material aspects of the scene.

The image has gone down in history as an illustration of "the Indians' victimization by the Spanish."[45] One detail is particularly eloquent in this respect:

One laborer, bending over to pick up extracted matter, shows us his buttocks. The image doesn't tell us how to read this detail, but at least two interpretations suggest themselves: (1) We might see here something anal and sodomitic—De Bry perhaps presents these indigenous buttocks for the European viewer's gaze as if mimicking the violent taking of minerals by these same Europeans—or (2) we might see here a gesture of revolt by the miner: "You want to look at me slave away in this mine for your empire, here instead are my buttocks for you to look at!" Whereas Figure 15 proclaims that Cerro Rico is a rich New World mountain whose mining potential has been discovered by the justice-bringing Spanish and which awaits European investors, Figure 16 responds: The Spanish are responsible for organizing large-scale mining at Cerro Rico, but that anthroturbation is forced upon and carried out *not* by *anthropoi* but by differentiated humans, by naked indigenous workers who labor in dangerous conditions, going up and down narrow ladders, carrying candles for which they are only partially reimbursed. De Bry, anticipating modern journalists, the pope, Naomi Klein, and thinkers such as Dipesh Chakrabarty, crafts an image that particularizes the abstract idea of humans-as-geological-force, capable of reshaping a mountain, by showing that such a project affects especially a specific, disenfranchised, population caught up in history and politics.

TERRAFORMATIONS

The discussion so far, of Las Casas, of images of Cerro Rico—and specifically their differences from each other—brings us back to Montaigne in order to grasp at the kind of thinking that he is pursuing in "Des coches," which likewise grounds its critique of European domination in the New World in a sustained, albeit somewhat backgrounded, reflection on what it means to take "stuff" from the ground. What makes Montaigne's chapter an interrogation of the exterranean in particular and not *just* a reflection on New World gold or colonial violence is that Montaigne opposes mining to another way in which we "take stuff from" and "reshape" the Earth. In a gesture whose origins can surely be traced to Ovid—as already at various points in Chapters 1 and 2 of the present book—Montaigne opposes, in his discussion of New World colonialism, different ways in which humans relate to Terra. It is as if he is saying: We *could* dig huge mines and extract gold, but we *might* establish other relationships with the planet, other kinds of habitability. What Montaigne brings back from the New World in "Des coches" is thus not gold but a countering sense of the exterranean. Both Niavis and Ronsard call upon their reader's familiarity with the shallow anthroturbation of agriculture, Niavis to set the latter in opposition to mining, Ronsard to blur the boundaries. Montaigne also contrasts extraction and farming, but in a remarkably different way—rather than with the mythical farming of Ovid's early ages, it is with a certain indigenous New World form of agricultural terrformation that Montaigne will establish a connection.

To think through how Montaigne's chapter counters the unrestrained European desire for huge-scale extraction, we must turn to its final pages. There, the essayist tells his reader about what was known to the Inca, as in Quechua today, as Capac Ñan, the twenty-five-thousand-mile road system created by the Incas, which linked—and still links, to some extent—the mountain peaks of South America to the continent's tropical lowlands. A road system might seem quite disconnected from the chapter's emphasis on mining and colonial violence, but resonances abound. This is how Montaigne begins:

> Quant à la pompe et magnificence, par où je suis entré en ce propos, ny Graece, ny Romme, ny Aegypte ne peut, soit en utilité, ou difficulté, ou noblesse, comparer aucun de ses ouvrages au chemin qui se voit au Peru, dressé par les Roys du pays, depuis la ville de Quito jusques à celle de Cusco (il y a trois cens lieues), droict, uny, large de vingt-cinq pas, pavé . . .

> (As for that ostentatious magnificence which led me to embark on this subject, neither Greece, nor Rome nor Egypt can compare any of their constructions, for difficulty or utility or nobility, with the highway to be seen in Peru, built by their kings from the city of Quito to the city of Cuzco— three hundred leagues, that is—dead-straight, level, twenty-five yards wide, paved . . .)[46]

Montaigne says more, but for now I pause here. As countless commentators have noted, the essayist praises the Inca road system as part of a comparison between the Old and the New Worlds. In particular, he asks his reader to think back to an earlier point in the chapter in which he had spoken, among other things, of Old World amphitheaters and Coliseum-like structures that housed games, gladiators, mock naval battles, and such like. This comparison is important, recognized as one of many in the chapter that testify to the cultural relativism of the *Essais*. But Montaigne's mention of the Capac Ñan has, in the critical history of the work, been swallowed up by the awareness for the comparison's relativizing power, as if all Montaigne had wanted to say was: Look, the New World is just like the Old World, its engineering projects are just as big and just as ostentatious. But if we bracket this comparison for a moment in order to direct our attention to the Inca road system itself, something else comes into focus. Regarding the Capac Ñan, Montaigne continues as follows:

[Le] chemin qui se voit au Peru [est] droict, uny, large de vingt-cinq pas, pavé, revestu de costé et d'autre de belles et hautes murailles, et le long d'icelles, par le dedans, deux ruisseaux perennes, bordez de beaux arbres qu'ils nomment molly. Où ils ont trouvé des montaignes et rochers, ils les ont taillez et applanis, et comblé les fondrieres de pierre et chaux. Au chef de chasque journée, il y a de beaux palais fournis de vivres, de vestements et d'armes, tant pour les voyageurs que pour les armées qui ont à y passer.

(The highway to be seen in Peru . . . is dead-straight, level, twenty-five yards wide, paved, furnished on either side with a revetment of high, beautiful walls along which there flow on the inside two streams which never run dry, bordered by those beautiful trees which they call *molly*. Whenever they came across mountains and cliffs they cut through them and flattened them, filling in whole valleys with chalk and stone. At the end of each day's march there are beauteous palaces furnished with victuals and clothing and weapons, both for troops and travelers who have to pass that way.)[47]

The focus here—to borrow the title of Jennifer Baichwal's film—is the manufacturing of landscape, that is, anthropogenic terraformation. In Montaigne's account, the Capac Ñan is an impressive feat of engineering not just because of its length, breadth, or the fact that mountains were "cut through" and "flattened" but because of its complexity: The road is paved; it is flanked by streams, trees, and walls. Of central importance here is that Inca engineers constructed this road system and gave it its specific shape and structure by drawing on knowledge garnered not in mining but in agriculture, especially as regards how to build and properly drain fields (or roads) on inclines and slopes. When Montaigne mentions the road's walls and streams, we must thus think not of decoration but of provisions specially set in place to dissipate the energy of water that would necessarily run down the hills, in order to protect the road from erosion.[48] In the words of archaeologists: "Following practices perfected in building walls to create terraced farm fields," the walls built along the Inca road system were constructed "with local stones and had open joints so that water could drain."[49] There is indeed a great proximity between agriculture and engineering. The stepping visible in the agricultural terraces in Figure 17 is also used, at a much smaller scale, to create a 10% incline in the paths used for walking. By mentioning the walls, the streams, and trees that border the road, Montaigne brings the kind of terraformation involved in road construction closer to agriculture than to that involved in mining.

The European invaders' searching out of mines, which would soon be turned from relatively artisanal into protoindustrial in terms of their output,

FIGURE 17. Agricultural terrain along the Capac Ñan. Werner Forman Archive / N. J. Saunders / HIP / Art Resource, NY.

is thus directly opposed to the indigenous cutting into mountains. It is not just that the road system is comparable to Roman amphitheaters in magnificence. Modes of terraformation are also set in opposition. This text most often recognized as inaugurating a new kind of cultural relativism, as critiquing European empire, as giving a voice to New World humans, thus also focuses on the nonhuman and on the different ways in which Old and New World humans apprehend the nonhuman in general and the exterranean in particular. Mining is identified as the primal fantasy of the colonizer, and it is mining that, in Montaigne as in Las Casas and De Bry, will lead to the violence exacted on New World inhabitants. But it is not just that mining is "bad"—or that it is the "cause" of colonialism—but that it implies a certain relationship to what goes from being sub- to exterranean, a relationship that can be opposed to other ways of cutting into Terra, for example, for the creation of a sustainable habitat like the Capac Ñan. In the way that Montaigne sets in opposition the European desire for large-scale extraction of gold and the manner in which the Incas engaged in large-scale Earthworks in the context not of mining but of infrastructure, Montaigne thus engages in something else: a reflection on the exterranean, on what it means to take a pickaxe to the Earth. The irony, of course—and Montaigne is nothing if not ironic—is that the Capac Ñan, cut from the Earth, would be transformed into the road system on which the newly arrived Europeans would transport their extracted matter.

To conclude, let us return to Montaigne's first mention of the Spanish as they sail "along the coasts *on the lookout for the natives' mines*" ("la mer *à la queste de leurs mines*.")[50]

Montaigne subsequently describes the landing of the Spanish in the Americas as well as the four declarations they make in addressing the native populations: (1) that, sent by the king of Spain and authorized by the pope who had given the New World to the Spanish, they come in peace; (2) that, if the indigenous populations submit to the Spanish king, they will be treated well; (3) that they want food and gold (for *medical* purposes!); and (4) that the indigenous populations would be well advised to believe in the Christian God.[51] Of the various responses that the Europeans hear is this one: "As for their being men of peace, if there were they did not look it."[52] The reversal of perspectives here—that is, a non-Spanish view of the conquistadors—stands as another emblem of Montaigne's cultural relativism, but if we switch back to the French, we might well see Montaigne making another key point that relies specifically on the French-ness of the expression: "Quand à estre paisibles, ils n'en portoient pas la mine," a literal translation of which might be: "As for being peaceful, they did not wear such a face [*la mine*]." In French, both sites of extraction and faces are called *mines*. The Spanish came searching for *mines* while claiming to be peaceful, but their faces (*mines*) reveal the truth: Their *mines* point toward *mines*. Whether this play on words was intentional on Montaigne's part, it captures the fact that Europeans and Native Americans do not see *mines* (and their relationship to Terra) the same way.[53] Unchecked extractivism and sustainable shepherding of landscape, even when huge-scale manufacturing thereof is deemed useful, are marked as *different* and as *different* along anthropological lines. And it is New World populations that suffer when extractivism is exported.

The "best thing about the Earth is if you poke holes in it oil and gas come out," said Republican congressman Steve Stockman in 2013.[54] The European explorers who set out to "discover" and colonize the New World likely had very similar thoughts—and a similarly instrumental and extractivist relationship to the Earth—when they dreamed about the gold and silver to be extracted in Hispaniola, Puerto Rico, the Kingdom of Yucatán, Potosí, and elsewhere. This is surely why Münster's *Cosmographia* speaks about metals and mining *before* even describing the size and divisions of Terra. At the moment that Terra is (quite literally) globalized, it is as a Terra ready for exterranean activity. Ronsard's Terre might belong more to mythology than to geography, but it is also a mythological translation of the globalized Terra of Münster and Montaigne's Spanish explorers. The critics of early modern Spanish colonialism studied here, De Bry, Las Casas, and Montaigne, implicitly reflect on the nature of the human relationship to Terra by putting the emphasis on the ways in which European-sponsored mining affected the indigenous populations of the New World. They were *mined*, they tell us, just as much as the continent was. Like Yann Arthus-Bertrand's documentary *Home* (2009), De Bry, Las Casas, and Montaigne offer "aerial shots" of different places on Earth that have been shaped by large-scale human practices, while simultaneously foregrounding the political and anthropological differences that attend to such anthroturbation.

PART II

WELCOME TO MINELAND

In the two chapters of this second section, to nourish further the notion of the exterranean, our attention shifts from (primarily) capital-*T Terra* to (primarily) small-*t terra*. This is not, at least not in any simple way, a switch from global to local—both because Part I showed the very idea of the global to be insufficient and because the local too has been traditionally associated with reactionary positions and with the notion of landscape.[1] In these two chapters, I want to bring us into contact with *terra*, with rocky materials, with earth, with dirt, with what we experience when we are physically present in a site of extraction, albeit via half-a-millennium-old texts and images. As Latour says of *le terrestre* (the terrestrial), the *terra* under examination in these two chapters is not "the frame of human action" but rather a "political-actor," and our aim—or at least our mandate—is indeed, as Latour sets out, as follows: "to know as coldly as possible the hot activity of a *terra* apprehended up close."[2] The point, again, is not to turn to what journalists would call local environmental problems but to ask what sense(s) of the exterranean become perceivable when we look at *terra* up close while still keeping in mind *terra*'s necessary connection to *Terra*. What connects *Terra* and *terra* is the (early modern sense of) *physis* or *natura*. In what follows, then, as we move to hillsides, mountains, and mines themselves, there will be much emphasis on detecting the vibrancy/the aliveness of *terra*—not as our ultimate purpose but as a step toward fashioning an alertness to how human extractors extract not from an Earth on which they stand but from a living Earth/*terra* into which they enter and which also enters them, in order to further nudge the exterranean away from visions of Earth seen from nowhere.[3]

This part of *Exterranean* inherits its name from a series of plays, films, and events organized by Philippe Quesne, director of the Théâtre Nanterre-Amandiers just outside of Paris, in the fall of 2016 under the title *Welcome*

FIGURE 18. Philippe Quesne's play *La nuit des taupes* (*The Night of the Moles*). © Théâtre Nanterre-Amandiers.

to Caveland! The rich program, which included inter alia Philippe Quesne's play *La nuit des taupes* (*The Night of the Moles*), Apichatpong Weerasethakul's "performance-projection" of *Fever Room*, and Cécile Fraysse's *L'île aux vers de terre* (*The Island of Earth Worms*), offered multiple perspectives on underground life, life forms, and living, as well as on Earth, the earthly, and the exterranean. As Marion Siéfert put it in the theater's official booklet, *Welcome to Caveland!* drew its inspiration "from the subsoil and from caves, from those *underground* and alternative territories that are close to the earth and to matter, but also conducive to dream visions."[4] Quesne's *The Night of the Moles*, in particular, strives to make perceptually available the material operation whereby animate beings penetrate into a *terra* that teems with life. The first thing that audiences see is the giant paw of a seven-foot-tall mole performing its exterranean activity by creating a hole in a giant white box in the center of the stage (Figure 18). For the rest of the play, moles dig tunnels, move matter about, play music, copulate, live, die, fight, and make art in underground spaces devoid of spoken human words. As I have written elsewhere, watching *The Night of the Moles* is "like getting a very satisfying injection of materialist philosophy cut with something vaguely psychotropic," leaving the spectator with the impression of understanding "Mole-Paw-Being."[5] The other plays and performances of *Welcome to Caveland!* similarly invite audiences to experience different aspects of living underground.

In a similar manner to *Welcome to Caveland!*—as self-invited additional moments within the festival[6]—the two chapters that follow also attempt to perceive exterranean activity *close up*. Chapter 4, "Sickly Mountainsides," offers a counter-reading of the first miner's bible, Georgius Agricola's *De re metallica* (1556). Usually read within the framework of a teleological narrative of technological progress, I here argue that Agricola's text also, despite itself, gives a voice to the nonhuman Earth. The stakes here are not parochial. Substituting one reading of a sixteenth-century text for another is not, in and of itself, necessarily urgent. But it becomes so, I believe, when the limitations of the reading to be substituted seem to have their origins in a bigger story, namely the modern constitution's sanctioning of *who* can say *what* (for example, Boyle can only speak of science, Hobbes of politics).[7] Chapter 5, "Demonic Mines," analyzes early modern texts and images that give voice to the nonhuman vibrancy of mines via the widespread belief in mining spirits. Emphasis will be on Agricola—his *De re metallica* but also his earlier *De animantibus subterraneis* (*On Subterranean Creatures*)—as well as on Paracelsus and the French author François Garrault. Just as Chapter 4 examines the livingness of extraction sites via close readings of landscapes and mountainsides, in order to infuse the exterranean with a close-up sense of *terra*, so Chapter 5 considers these mining spirits for the way they animate the sites of extraction and the chemical materialities that attend the removal of matter.

CHAPTER 4

SICKLY MOUNTAINSIDES

The twelve books of Georgius Agricola's *De re metallica* (*On Metals*) first published in Latin in 1556 detail almost all aspects of extraction and, as such, capture not only how early modern miners took matter *ex terra* (in which places, with what tools, at what depths, against what opposition, etc.) but also how that process was phenomenalized as part of a wider ecology linking *Terra*, sites of extraction, *terra*, and extracted matter.[1] Subsequent German and Italian editions of the *De re metallica* quickly entered circulation, making the book—for a significant amount of time—"the miner's bible" of Europe.[2] Book 3 maps out whole underground worlds by delineating the different kinds of veins; Book 4 offers in-depth explanations of how to measure, mark out, and assign ownership to underground deposits; Book 5 looks at shafts, tunnels, and digging, as well as the art of surveying; and Book 6 analyzes iron tools (hammers, crowbars, picks, buckets, barrows, etc.) and machinery (for hauling and ventilating), etc. The final books (7–12) describe how to process extracted matter (assaying, roasting, smelting, separating, etc.).[3] According to generations of scholars, it is primarily a technical book.[4] Given the preceding summary, such a judgment is not unjustified. The fact that Book 1 is, as we shall see below, a long defense of extraction further adds to such a reading, according to which Agricola sets up a relatively cut-and-dry relationship between humans and matter and in which human agents and their manmade tools extract matter from a submissive planet, as if theirs to mine as they see fit. But such a reading is part of a much bigger story. If we have understood the *De re metallica* as an exclusively technical book up until this point, it is largely because of what Latour calls the modern constitution; that is, just as Robert Boyle came to be read exclusively as a scientist, and just as his contemporary Thomas Hobbes was allowed to be heard only on politics, so Agricola's writings have traditionally been received only for what they say on the technical aspects of mining.[5] The stakes are high. In what follows, I shall argue that if we allow the *De re metallica* to exit the purely technical domain, it becomes quite a different work. What follows is thus a counter-reading of the *De re metallica*, one that will plot moments in the work that can contribute to a sense of the exterranean seen up close, at the scale of the mountainside and the shafts and tunnels sunk into it. Something else, we shall see, becomes perceptible, namely an awareness for the livingness of the mountainside's surfaces and innards and consequently for a sense of how extraction sites are actors in the production of exterranean matter and of the exterranean as a conceptual category. As Latourian as this enterprise is, it also rejects Latour's point in his recent *Où atterrir?* (*Where to Land?*) that the Anthropocene has no precedent. Latour's presentation of the Holocene in that work—as having "all the characteristics of a 'framework' inside which it is possible, without too much difficulty, to distinguish human action—just as at the theater one can forget the building and all that is offstage and focus only on the plot"[6]—clashes quite strikingly, as we shall see, with the way in which Agricola depicts exterranean activity in the *De re metallica*.

GEORGIUS AGRICOLA

As this will be the first extensive discussion of Georgius Agricola (1494–1555) in this book, it is worth pausing to introduce him and his work. After mastering both Latin and Greek in Germany, reading and admiring the works of Erasmus, generally receiving a humanist education, and working as a member of the editorial staff for the Aldine editions of Galen and Hippocrates in Italy, Georgius Agricola got married and settled down to begin a quiet but busy life as a paterfamilias, physician, mineralogist, investor in mining, and writer in the town of St. Joachimsthal, now Jáchymov in the Czech Republic.[7] This was in 1527. He moved there in particular as a more or less direct consequence of something that had happened eleven years earlier. The general area had long been a mining region. But in 1516 the region's prominence suddenly grew when the prospector Pfandherr Stefan Schlick struck lucky. Very quickly, a mining industry grew up in the town, inaugurating a period of rapid financial, demographic, and environmental change.[8] Major banks and businesses invested money in the region, such that the hitherto largely uninhabited hills of the Erzgebirge were suddenly overrun with miners, pit foremen, master builders, and stone grinders, as well as with craftsmen of all types, from sculptors and glaziers to painters, jewelers, and silver- and goldsmiths. By 1525, the "Free Mining Town" of St. Joachimsthal had a population of 13,411. As well as spending much of his time visiting mines and smelting houses and treating patients with various (often mining-related) ailments, Agricola wrote a number of works, all of which entertain a more or less direct relationship to this proximity to and financial and intellectual interest in extraction. He wrote geological and mineralogical treatises that greatly extended technical and conceptual mastery over planetary matter.[9] And he also wrote about—and indeed spilled much ink in rabid defense of—mining, first in his *Bermannus, sive de re metallica* (1530), a humanist dialogue between a mining expert of Joachimsthal and two scholar-physicians, a work already briefly mentioned in Chapter 2,[10] and second in the *De re metallica*, begun in 1533 but published only posthumously. Around the same time that Ramus in France was promoting the idea that education needed to be *useful*, that intellectual training be linked to practical applications, Agricola set about using *his* humanist training in ancient languages and philology to promote human mastery of the physical planet in order to maximize profits.[11] But the sense of the exterranean that his works develop is arguably less clear-cut.

A (NOT WHOLLY SUCCESSFUL) DEFENSE OF EXTRACTION

Let us turn first to Book 1 of the *De re metallica*, which is, in the rapidest of summaries, a "defense of the profession" of mining.[12] As another reader put it, Agricola, aware that "mining did not always enjoy social acceptance as a trade," set out to defend it "for reasons of justice and on the basis of moral principle."[13] But Book 1 is also more (and, as we shall see, less) than a defense. It certainly attempts to be a defense in ways that go beyond arguing for "social acceptance" of a "trade," for its arguments have much more to do with the ecological than the social and much more to do with mining defined as an art rather than as a mere trade. This is an important first point, as it already suggests that the *De re metallica* sees itself as a work whose focus is not as single-minded as its reception history suggests. What Agricola attempts in his apology is a naturalization of extraction, ultimately leading to the comparison between extracting matter *ex terra* and the taking of fish from the sea: "It is far stranger that man, a terrestrial animal [*hominis terreni animalis*]," he writes, "should search the interior of the sea [rather] than the bowels of the earth [*terrae viscera*],"[14] a sentence whose rhetorical power lies in its assertion that mining is *more* natural than an activity likely assumed by his readers to be wholly uncontroversial. The exterranean is cast as a particularly human shifting of matter but also as an activity that takes part within wider sets of ecological connections. Moreover, Book 1, as much as it is a defense and indeed as it tries to be a defense, also gives plenty of space to those other humans who disagree with Agricola and who see its negative effects. Before exploring the intricacies of what and how Agricola advances his argument here, it is first worth recalling what he says to rebut his opponents' (supposedly) most substantial claim against him:

The strongest argument of the detractors is that the fields are devastated [*agri vastantur*] by mining operations, for which reason formerly Italians were warned by law that no one should dig the earth for metals [*terram foderet*] and so injure [*corrumperet*] their very fertile lands, their vineyards, and their olive groves. Also they argue that the woods and groves are cut down, for there is need of an endless amount of wood for timbers, machines, and the smelting of metals. And when the woods and groves are felled, then are exterminated [*exterminantur*] the beasts and birds, very many of which furnish a pleasant and agreeable food for man. Further, when the ores are washed the [used] water poisons the brooks and streams, and either destroys the fish or drives them away [*aut necat, aut ex eis abigit*]. Therefore the inhabitants of these regions [*incolae regionum*], on account of the devastation of their fields, woods, groves, brooks and rivers, find great difficulty in procuring the necessaries of life, and by reason of the destruction of the timber they are forced to greater expense in erecting buildings. Thus it is said, it is clear to all that there is greater detriment from mining [*plus in fossione detrimenti esse*] than the value of the metals which the mining produces.[15]

Agricola's detractors, even in his summary of them, do a good job of showing the extent of the impact of mining on the area around the site of extraction: Mining ruins farmland, kills animals (including those to be consumed by humans), pollutes the water supply, and makes it difficult to procure even the basic necessities of life. Given that Agricola is ostensibly trying to defend mining, the precision with which he quotes his adversaries is striking, for it brings them into this printed archive in a way that was otherwise unlikely to happen. From these accusations what emerges is not *just* a trace of so-called local environmental degradation, for which a local solution might arguably be found, but more compellingly a sense of the entanglement of activities, habitats, and human and nonhuman beings and thus of how the exterranean is not just a taking from the Earth, that is, an activity involving an origin and an action, but an activity that affects—and jeopardizes the livingness of—that origin in multiple and manifold ways. The price of extracting *ex terra*, according to Agricola's opponents, is the destruction of that point of origin. Agricola's specific response to this set of charges obeys no logic other than that of Freud's kettle logic,[16] in which arguments are multiplied to the point of becoming inconsistent with one another:

As the miners dig almost exclusively in mountains otherwise unproductive [*montes nihil frugum ferentes*], and in valleys invested in gloom [*valles tenebris circumfusas*], they do either slight damage [*vastitatem exiguam*] to the fields or none at all [*aut nullam*]. Lastly, where woods and glades are cut down, they may be sown with grain after they have been cleared from the roots of shrubs and trees. These new fields soon produce rich crops, so that they repair the losses which the inhabitants suffer from increased cost of timber. Moreover, with the metals which are melted from the ore, birds without number, edible beasts and fish can be purchased elsewhere and brought to these mountain regions.[17]

Following Agricola's kettle logic, then: There is no devastation, because mining only happens in infertile regions; there is only slight damage; there is damage, but it can be repaired relatively easily; there are losses, but soon these areas will produce rich crops; mining generates money, which means animals can be bought elsewhere and brought to the devastated areas. It is hard to see this as anything other than a scramble for responses where there are none. Other accusations and rebuttals are just as muddled and ungrounded. At one point, for example, Agricola quotes his detractors, who seemingly look to Socrates for an authority who agrees with their point of view: "It is said that Socrates, having received twenty *minae* sent to him by Aristippus, a grateful disciple, refused them and sent them back to him by the command of his conscience. Aristippus, following his example in this matter, despised gold [*aurum sprevit*] and regarded it as of no value [*ac pro nihilo putavit*]."[18] A number of pages later, Agricola rebuts by asserting rather that "Socrates, in truth, did not despise gold [*vero non sprevit aurum*]"—if he refused gold, it was merely because he "would not accept money for his teaching."[19] Wherever one dips into Agricola's so-called defense, one comes upon similar back-and-forth exchanges, in which Agricola seems at a loss to justify his own position. He clearly *wants* to legitimate extraction and to naturalize the exterranean, but it is hard not to hear the detractors' voices much more loudly.

To get a fuller sense of how Agricola's defense works, let us examine how, by defining mining not as a trade but as an art for which one must receive a humanist education and by engaging with the humanist practice of dialoguing with the ancients, Agricola attempts to attach his reevaluation of mining to some of the discussions already seen in previous chapters. First, then, we should recall that, as Agricola summarizes in his preface, Book I contains the arguments that "may be used against this art [*hanc artem*], and against metals and the mines, and what can be said in their favor."[20] Agricola's reference to the fact that he will be setting out arguments for and against mining *as an art* (Latin *ars*) is essential, for the first section of Book I will define in some detail the desired education for mining professionals (miners but also mine managers, etc.). In other words, and similarly to how Vitruvius defines the architect's education in his *De architectura* (first printed in 1486), the miner must not just be a laborer who receives instructions but an educated humanist in his own right.[21] Just as Vitruvius begins by asserting that the good architect "should be equipped with knowledge of many branches of study and varied kinds of learning" and that he should privilege a form a "knowledge [that is] the child of practice *and* theory," so Agricola claims that "there are many *arts and sciences* of which a miner should not be ignorant."[22]

From the outset, Book 1 announces extraction as an activity that is more than merely technical in the narrow sense, for its human agents must be trained in various disciplines: Medicine allows the mine manager to "look after his diggers";[23] astronomy teaches the miner how to "judge the direction of the veins";[24] surveying enables him "to estimate how deep a shaft should be sunk to reach a tunnel" and also how "to determine the limits and boundaries in these workings";[25] math serves "to calculate the cost to be incurred in the machinery and the workings of the mine."[26] Also useful is architecture, for constructing "the various machines and timber work required underground" and drawing the sketches that allow for the preparation of "plans of [the miner's] machinery."[27] And then there is law (*jus*), useful for "claim[ing] rights" over mined metals and also to make sure that the miner does not "take another man's property," which would "make trouble for himself."[28] Of particular note for our purposes, however, is not just the *making-useful* of established disciplines—in a way that echoes the same turn (in Anthony Grafton's terms) from *humanism* to the *humanities* exemplified by Ramus in France—but also the fact that the very first of the disciplines that Agricola lists, even before *medicina*, is *philosophia*. Philosophy, argues Agricola in a juxtaposition that stresses the leap between scholarship and practically applicable science, will allow the miner both (1) to discern "the origin, cause, and nature of subterranean things" and (2) "to dig out the veins easily and advantageously," in order to "obtain more abundant results from his mining."[29]

Ironically, it is precisely just this miner's education that allows Agricola to provide the reader with a veritable cento of texts that his opponents usually quote, including Euripides, Phocylides, and many others. As we saw in Chapter 2, both Ronsard and Agricola quote (and refute) Naumachius's claim, via Stobaeus, that "gold and silver are nothing more than dust, like the stones that can be found on the pebbly beach, or else on the banks of rivers."[30] But that is just one of many. The first and clearly the most important of all the authors whom Agricola quotes—in order to refute—is Ovid. Indeed, although the "strongest" arguments of his detractors are those that have to do with the devastation of farmlands, water pollution, and suchlike, Agricola begins his discussion of "those critics who say that mining is not useful" by aligning their position, generally writ, as a reading of Ovid's four ages, as defined in the *Metamorphoses* and that we have already seen to be central to discussions on mining, for example in the *Judicium Jovis* discussed in Chapter 1. They argue, he says, that the "Earth does not conceal [*Terra non occultat*] and remove from our eyes those things which are useful and necessary to mankind"; rather, Earth, "like a beneficent and kindly mother [*mater*]," yields "in large abundance" all things, especially "herbs, vegetables, grains, and fruits, and the trees."[31] They argue, too, that Earth buries metals and minerals "far beneath in the depth of the ground" ("fosilia in profundo penitus abstrudit"), an opinion that Agricola attaches to Ovid, who, he says, "censures their audacity" ("eam audaciam merito insequitur"). At this point, Agricola quotes these lines from Book 1 of the *Metamorphoses* about the iron age:

Nec tantum segetes alimentaque debita dives
poscebatur humus, sed itum est in viscera terrae,
quasque recondiderat Stygiisque admoverat umbris,
effodiuntur opes, inritamenta malorum.
iamque nocens ferrum ferroque nocentius aurum
prodierat, prodit bellum.

(Not only did men demand of the bounteous fields the crops and sustenance they owed, but they delved as well into the very bowels of the earth; and the wealth which the creator had hidden away and buried deep amidst the Stygian shades, was brought to light, wealth that pricks men on to crime. And now baneful iron had come, and gold more baneful than iron; war came.)[32]

For Agricola's critics, extraction *ex terra* is really a movement of matter not just from the "bowels" of a mother but also from the infernal and dark regions of the Pagan underworld. It is an activity that, after Ovid, brings iron, gold, and also war into the world. Both the specific details and general flow of this argument are familiar from Chapter 1. Agricola does not respond directly to his opponents' reading of Ovid—he produces the quotation and moves on rapidly to enumerate his opponents' other claims (*they rebuke, they blame, they clamor*, etc.), resulting in a cacophony of quotes with which he ostensibly disagrees. Only a number of pages later does Agricola respond to the contrary opinion supported by the citation from Ovid. He writes—with no ancient authorities to back him up, moreover—as follows: "The Earth does not conceal metals in her depths because she does not wish that men should dig them out, but because provident and sagacious Nature has appointed for each thing its place"—an argument that we have already seen advanced by the miner in the *Judicium Jovis*.[33] Agricola provides further explanation of his point: Nature "generates [metals] in the veins, stringers, and seams in the rocks, as though in special vessels and receptacles for such material."³⁴ If, argues Agricola, metals have their proper "abiding place in the bowels of the earth," then who would not agree "that these men do not reach their conclusion by good logic?"³⁵ It is at this point, as a final refutation, that Agricola makes his argument about extraction *ex terra* being like taking fish *from the sea*.³⁶ Agricola appeals then to Natura—and also, subsequently, to the Christian God: "Those who speak ill of the metals and refuse to make use of them, do not see that they accuse and condemn as wicked the Creator Himself."³⁷

The debates over the usefulness of extraction in Book 1 of the *De re metallica* deal in some senses with a local problem: the deforestation and destruction of agricultural land in the mountains that separate Saxony and Bohemia. But the arguments that go back and forth are *not* local or localist arguments. Rather they betray what Mitchell Thomashow has said much more recently, that there "is no such thing as a local environmental problem."³⁸ Both Agricola's detractors and Agricola attempt to shape the discussion on the disruption of *terra* by relating it to the way in which *Terra* more generally is conceptualized and regulated. Art and science, including the tools of philology, are brought to bear on the question. Rather than in the domain of the local, we are in that of the Latourian terrestrial: "The *terrestrial* [*terrestre*] coheres with the earth [*la terre*] and with ground [*le sol*], but it is also worldly [*mondial*] in this sense that it coincides with no frontier and that it overflows all identities."³⁹ By confronting Ovid with the concrete hypothesis that Nature places ore in specially designed underground veins, by arguing over whether mining announces a new age or not, Book 1 does the opposite of what Thomashow and certain modern ecological thinkers do—rather than saying that a local issue affects Terra on a global scale, Agricola tries to argue that the way in which Terra, Natura, and the Christian God place ore in veins organizes *and condones* any perceived local disruption. The sense of the exterranean seen up close has much more to do with a collapsing of scales, then, than with a change of scales.

SOME (RATHER SAD) MOUNTAINSIDES

The original 1556 edition of the *De re metallica* includes 292 richly wrought woodcuts, based on original drawings by Blasius Weffring and executed by a team of woodcutters, including two whose names we know (Hans Rudolf Manuel Deutsch and Zacharias Specklin), which offer illustrations of hillsides, shafts, veins, tunnels, miners, machines, and many other technical and quotidian details of extraction.⁴⁰ To anyone consulting an original edition of the work, even a digitized version, it will be clear that their presence is anything but decorative—they not only make the technical details explained in the text immediately visible and, one assumes, instantly applicable, but they also communicate a more general sense of how humans, machines, mountainsides, and matter interact with one another. The full title of the *De re metallica* emphasizes how crucial the woodcuts are to the project as a whole—they are not an afterthought or a marketing tactic: *On Mining, in Twelve Books, in which the Operations, Instruments, and Machines, as well as Everything That Pertains to Mining are not only clearly described* [non modo luculentissime describuntur] *but also set before the reader's eyes* [ob oculos ponuntur] *in the form of pictures* [per effigies] *inserted at the appropriate places with their Latin and Greek names.*[41] Agricola expands that claim in his Introduction, stating that regarding the veins, tools, vessels, sluices, machines, and furnaces, he "not only described them" ("non modo descripsi") in his text but also "hired illustrators to delineate their forms" ("sed etiam mercede conduxi pictores ad earum effigies exprimendas") not just for the purpose of decoration but more specifically in case "descriptions which are conveyed by words [*quae verbis significantur*] should either not be understood by men of our own times, or should cause difficulty to posterity."[42] They communicate many technical details with striking clarity—and this much has been acknowledged by generations of critics, as the wider reception history would lead us to expect. But again, if we allow the *De re metallica* to escape a purely technical or technological framework, thus freeing ourselves to look at the woodcuts with fresher eyes, things look different. I want to suggest that the images also contain a certain excess—when compared to the texts that they accompany—that makes reading them more complicated and their sense of the exterranean much richer. More specifically, I want to explore if—and how—we might read the woodcuts as expressing precisely the anxiety over extraction, at the local level, that Book 1, as we saw above, seeks to repress.

The woodcuts included in the *De re metallica* have been read in a number of ways. Pamela Smith has demonstrated convincingly how Agricola's images belong to a new early modern visual culture—also visible in, for example, the herbarium of Leonhart Fuchs or the anatomical treatises of Vesalius—in which strict distinctions between art and science and between making and knowing do not obtain, with artisans playing a key role in the production and communication of technical knowledge and more generally knowledge about the physical world.[43] However, this connection between the images in the *De re metallica* and other contemporary works, which in part explains their lifelike and naturalistic quality, is not sufficient to account for how we should read one particular detail in many of the woodcuts, namely their almost constant emphasis on the degradation caused by mining. Take the hillside shown in Figure 19, from a section of Book 5. It shows a hillside into which three shafts have been sunk as well as a tunnel in which miners are already working. It also shows the construction of wooden "shaft houses" above the shafts, whose purpose is to protect against one of the greatest dangers of mining, the flooding of the underground galleries. This much is technical. But the image also features, scattered throughout it, tree stumps, which identify where trees have been cut down, precisely to provide the wood needed for creating and protecting the shafts. Questions abound. Are these stumps shown merely to make the image lifelike and naturalistic? Or do they do something else? Might they speak, rather, to the sense of devastation that causes Agricola's detractors to quote Ovid? Are they signs that extraction *ex terra* is more antagonistic than Agricola would have us believe? The commentary of one recent reader, Owen Hannaway, is purely descriptive, avoiding such questions: The picture shows, he writes, "the extensive deforestation that is taking place in conjunction with mining operations," adding that mining was "disruptive to the use of the land."[44]

Hannaway's comments on other images in the *De re metallica* are similarly descriptive rather than fully analytical. Thus, regarding a woodcut in Book 6 focused on the production of wooden pipes via the boring out of trees (Figure 20), Hannaway writes: "One of the reasons for the extensive deforestation in mining areas is the increase in the size and the complexity of the pumps and machinery."[45] In other words, the representation of felled trees might be read, according to Hannaway, as (nothing more than) a sign of (the presence of) technological progress. The author of the fullest study of these woodcuts to date, Marie-Claude Déprez-Masson, goes a bit further in her analysis of such representations. She divides details into those that are "justified" and those that are seemingly extraneous, for which she either frankly admits to have no explanation ("je n'ai pas d'explication") or explains as being there only as "charming snapshots of daily life" in the mines.[46] Visual details of deforestation, she says, belong to the first category of "justified" (textually motivated) representations. She thus agrees with Hannaway when she writes that landscapes "studded with tree stumps" and that "recall the intense deforestation linked to mining activities," as well as the often "insubstantial appearance" of trees in the foreground, which contrast with the relatively healthy ones in the background, serve to illustrate a piece of information that Agricola details in his text, namely how this specific kind of (lack of) vegetation serves as a sign allowing the reader to identify the area as an extraction site.[47] In this reading, then, the depiction of both (a) deforestation and (b) somewhat sickly trees in the foreground do not relate to the discourse of devastation explored in Book 1 but serve instead to identify an extraction site. Again, local devastation is framed in nonlocal terms, and the exterranean is, when observed in the extraction site, never fully contained by it.

FIGURE 19. Agricola, *De re metallica* (1556), 72. With kind permission of the Rare Book and Manuscripts Library, Columbia University.

FIGURE 20. Agricola, *De re metallica* (1556), 135. With kind permission of the Rare Book and Manuscripts Library, Columbia University.

Book 2 of the *De re metallica* gives much credence to such a way of reading the images. There, Agricola offers details about how to identify the best places for extraction—and much of Agricola's language turns precisely on the vegetative livingness, fertility, or lack thereof, of the sites; that is, the local is defined in terms of how Nature has distributed her own vitality differently across the whole of Terra's expanse. The question is indeed not simple. On one hand, Agricola argues that a miner should try to obtain a mine "to which access is not difficult, in a mountainous region, gently sloping, wooded [*sylvestri*], healthy [*salubri*], safe [*tuto*], and not far distant from a river or stream by means of which he may convey his mining products to be washed and smelted."[48] The site must thus be wooded precisely so that the trees can be cut down and turned into construction timber: "If it be a wooded place [*Si nemorosus est*], he who digs there has this advantage, besides others, that there will be an abundant supply of wood for his underground timbering, his machinery, buildings, smelting, and other necessities." And conversely: "If there is no forest [*Sin careat nemoribus*] he should not mine there unless there is a river near, by which he can carry down the timber [*ligna*]."[49] So, on this first hand, Agricola explains that the ideal extraction site is wooded—and thus condemned to being deforested.

On the other hand, the ideal extraction site is also somewhat less healthy (and less wooded?) from the very beginning. The absence of farming is, here, a positive sign: "A region which is abundant in mining products very often yields no agricultural produce [*nullas ferant fruges*]," such that the "necessaries of life for the workmen and others must all be imported."[50] Moreover, even plants that do grow in mining areas are, says Agricola, likely to appear less healthy: The presence of veins produces excessively hot exhalations, which can cause the foliage of trees to have a "bluish or leaden tint."[51] And more generally, the veins cause "the earth [to] produce only small and pale-colored plants" ("ea terra fert herbas humiles et coloris non vivi").[52] These exhalations can further scorch the roots of trees, which can render the roots and trees sickly ("infirmas reddunt").[53] A further detail is as follows: "In a place where there is a multitude of trees, if a long row of them at an unusual time lose their verdure [*amittunt viriditatem*] and become black or discolored [*nigrescunt, aut discolorantur*], and frequently fall by the violence of the wind, beneath this spot there is a vein [*ibi subest vena*]."[54] Thus, reading across Book 2, Agricola suggests that a miner should seek out an area that is "wooded," "healthy," and "safe" by (strangely) looking for an area that is wooded yet devoid of agricultural crops and populated by somewhat "small" plants and "sickly" trees. This being so, Déprez-Masson's point that woodcut representations of deforestation and insubstantial trees serve to designate and localize extraction sites seems correct. But *what*, in all this, of Book 1's attempt to minimize the connection between mining and surface devastation? Are not the woodcuts in their lifelike and repeated depiction of deforestation and sickly trees revealing that in fact Agricola's opponents were right? Do they not in fact testify to the extent of the damage caused by mining?

There is absolutely no way to know anything *precise* about the intentionality of Weffring, on whose drawings the woodcuts are based, or about what Agricola did or did not instruct him to depict, or still about the woodcutters themselves and the extent to which they directly followed Weffring or Agricola. As we just saw, it would be possible to justify the presence of the tree stumps and sickly trees by referring to Book 2 of the *De re metallica*. But what if we refuse being content and complicit with such an explanation? What if we decide—as an act of seeing and reading—to stop seeing sickly trees as (only) signs planted by Nature to indicate the presence of underground veins? What if we see the tree stumps not just as neutral realistic depictions of deforested land *because it is deforested* in order really to *see* that deforestation? Such a turn—antihistoricist, for sure—is the kind of turn defended by Michel Serres in his *The Natural Contract*. In the opening pages of that book, Serres offers a counter- and antihistoricist reading of Francisco Goya's painting *Fight with Cudgels* (*Riña a garrotazos*, or *Duelo a garrotazos*) (Figure 21). Serres suggests that, generally speaking, when we look at the painting we focus on the two humans who fight each other with sticks. We might, in betting mode, put our money on the man on the left or on the man on the right. But Serres argues that there is a third entity on which we might also choose to place our bet: "the marsh into which [the two men] are sinking."[55] As we look on at the pugnacious human subjects interlocked in battle, do we not, asks Serres, forget "the world of things themselves, the sand, the water, the mud, the reeds of the marsh?"[56] Serres extrapolates from this reading of Goya to suggest that in our current times of global warming—he would have likely said *in the Anthropocene*, had the word been in circulation when he wrote *The Natural Contract*—"What was once local—this river, that swamp—is now global: Planet Earth."[57] If we make the choice to set aside the question of intention and to read Agricola's woodcuts in a manner consistent with Serres's reading of Goya, then might we not indeed find that they connect back as much—or even more—with Agricola's detractors from Book 1 than with his own project of producing a technical how-to book about mining?

As an experiment in such a manner of seeing, let us turn to the very first woodcut in the *De re metallica*, which appears at the end of Book 2 in the midst of a long and detailed discussion of dowsing, that is, the use of a divining rod or *virgula* to locate buried metals (Figure 22). Guided by the accompanying text, the reader sees two different men holding Y-shaped twigs (labeled "A"),

FIGURE 21. Francisco Goya, *Fight with Cudgels* (*Riña a garrotazos*, or *Duelo a garrotazos*) (1819–1823). Museo del Prado / Art Resource, NY.

who seem to have mastered the technique, given that other men are now busy digging and seemingly extracting matter *ex terra*. The text provides all sorts of complementary details: Some miners swear by twigs torn from hazel trees; others prefer to use a specific kind of twig depending on which metal they are searching for (hazel for silver, ash for copper, pitch pine for lead and tin, and rods of iron and steel for gold). Agricola explains how prospectors use the rods in the most concrete and quotidian detail: "All alike grasp the forks of the twig with their hands, clenching their fists, it being necessary that the clenched fingers should be held toward the sky in order that the twig should be raised at that end where the two braches meet." Prospectors then wander hither and thither until the twig "turns and twists," thus "disclos[ing] the vein."[58] The *virgula* thus functions—if indeed it *does* function, for Agricola has his doubts, as we shall see—to connect the human miner and the nonhuman underground deposits.

The very practical nature of the text runs up against the humanist impulse to seek out classical connections. On the practical register, Agricola argues that, on the one hand, "we should not press the fingers together too lightly [on the rod], nor clench them too firmly, for if the twig is held lightly they say that it will fall before the force of the vein can turn it"; on the other hand, he warns that if "it is grasped too firmly the force of the hands resists the force of the veins and counteracts it."[59] Such highly practical and use-focused discussion of the divining rod, which allows us to imagine the curious physician asking questions far and wide of the men he meets in the Erzgebirge, gives way in due course to more philological and scholarly concerns, namely the correspondence (or lack thereof) between such usage of the *virgula* and the

Virgula A. Foſſa B.

FIGURE 22. Mining on a hillside, from Agricola, *De re metallica (1556)*. With kind permission of the Rare Book and Manuscripts Library, Columbia University.

opinions of the Bible and the ancients. As if to endorse the opinions of local artisans and laborers, Agricola adds that "in Homer, Minerva with a divining rod turned the aged Ulysses suddenly into a youth, and then restored him back again to old age." After several more examples, Agricola concludes, with a naïveté that it is difficult not to relish, "therefore, it seems that the divining rod passed to the mines from its impure origin with the magicians."[60] There has been, thus, a *translatio* of the *virgula* from ancient Pagan literature into the early modern mines. If we allow the text to guide our reading, then the woodcut in Figure 22 serves to show (1) how to use a divining rod and (2) the shift of this instrument from the domain of mythology and magic to that of the extraction industry.[61] Whether or not the *virgula* is efficacious is, however, in doubt: "Some say that [the rod] is of the greatest use in discovering veins, and others deny it."[62] Agricola's own opinion is that rods are actually of no use whatsoever. The miner, because Agricola esteems him to be "a good and serious man" ("virum bonum et gravem"), should not, then, "make use of the enchanted twig [*virgula incantata*]"—rather, he should be "prudent and skilled in the natural signs," for "there are the natural indications of the veins which he can see for himself without the help of twigs."[63] Use of the *virgula* is thus opposed to the identification and interpretation of *naturalia signa*. It is thus not the *virgula* that the reader needs to *see* in the woodcut but the natural signs connected with extraction.

Such, then, is what the text asks us to see in the image. If we choose to follow Serres's process, and, better to do this, if we listen to Jane Bennett's invitation to "cultivate the ability to discern nonhuman vitality, to become perceptually open to it,"[64] then the idea of considering the hillside only from the point of view of the prospector who seeks *naturalia signa* in order to practice exterranean extraction gives way to other perceptions. If we ignore the text momentarily, what we see is a hillside ripped apart—and indeed *being* ripped apart. The woodcut (Figure 22) shows, cartoon-like, different stages of the process: We see one man (top right) at the very moment when his pick is gathering momentum to pass back over his head and fall into the ground, we see (midground) a man whose pick is at the end of its swing trajectory, and in the foreground another man who has put his tools down to inspect extracted matter. But we see not only human agents. We see also numerous trees that have been cut down to mere stumps and that populate the woodcut like hairy moles on an aged face; we see (back center) a tree that seems to cower and lean away from a tool-wielding human intent on cutting off its branches. And we see destruction that avoids the binary human/nonhuman exchange: In the top-right corner, a fire burns, and smoke billows up in a space that would have previously been filled with foliage.[65]

There is also something—quite a lot, in fact—to be said of the woodcut's pointing hands. In the extreme foreground (left), we see two men with gloves removed and whose rigid index fingers point directly at the *A* and *B* legend letters, explained above the frame as standing for *virgula* (divining rod) and *fossa* (hole/trench). This direct lining up performs an erasure of the representational levels, between the diegetic scene (*miners on hill*) from the extradiegetic typographical convention of indexal letters, between the ecomimetic and the purely typographical, indeed between image and text.[66] Moreover, and this point is both historical and theoretical, for anyone familiar with the practices of early modern book readers, each of these two human hands strikingly resembles a manicule, that hand-with-pointing-finger symbol (some variation of ☞) that was probably, in William Sherman's words, humanism's "most common symbol . . . in the margins of manuscripts and printed books."[67]

The fact that these human hands are also typographical symbols, which point to other typographical symbols (A and B), which direct the reader to the legend's explanations (*virgula, fossa*), which in turn name and thus point back into the image, identifying both (A) a tool that itself points from man to hillside (and/or vice versa) and (B) an anthropogenic change in the hillside (the trench), serves to connect together all the humans and nonhumans involved in the extraction process. The hillside becomes, like a book, readable—not only because of *naturalia signa* but via this human and nonhuman network. Although seemingly nothing more than a technical "how-to" illustration of the use of the *virgula*, the woodcut *does* much more, telling the story of the progressive reshaping, or destruction, of the hillside.

Thinking more about both the manicule-hands-that-point and about the *virgula* via Heidegger's notion of *Zuhandenheit* (normally translated as "readiness-to-hand"), as indeed Sherman encourages us to do when he makes this connection in his history of the manicule,[68] would take us even further toward realizing the extent to which this supposedly "how-to" image *thinks*. In Raymond Tallis's account of *Zuhandenheit*, referenced by Sherman, Heidegger uses the term to denote "one of the fundamental, indeed primordial, categories of Being": "The world is composed primarily of 'handy being'; and 'handiness' is central to [Heidegger's] so-called 'existential analytic.' The world . . . far from being the traditional collection of 'objective presences' that constitutes the physical universe of science, facing the equally traditional isolate subject, is a nexus of 'the ready-to-hand' disclosed in, by and to Da-sein or 'being-there.'" The world—or here, the hillside—is thus "not a rubble-heap of matter" but "a network of meanings embodied in the ready-to-hand."[69] The woodcut shows the manufacturing of an extraction landscape, translates a powerful geoaffect, and refuses to situate, simply and unproblematically, humans *in an environment*, inviting its reader instead to pay attention to the hillside's cowering and forlorn trees and not just to the miners who successfully master usage of the *virgula*. We see not an environment but an ecological perception of the exterranean in action.

If we refuse to read the *De re metallica* as a purely technical work, as has been the choice made by most earlier readers in accordance, I would argue, with the modern constitution as Latour defines it, then what it offers us is a sense of the exterranean in which human agents extract not only from a *Terra* on which they stand or whose systems they understand (or not) in general but also from a *terra* into which they enter, which actively resists them, and which indeed at times *enters them*, making humans the passive receivers of *its* actions. Had the *De re metallica* not been read as a purely technical manual, had its awareness for the liveliness and corresponding potential deadness of hillsides and mineshafts not been forgotten or erased, had the belief in mining spirits not been put into parentheses as pure fiction, it is very hard to see how the COP21 agreement discussed in this book's Introduction could have spoken only of emissions and never of extraction. Some remnant of Agricola's close-up sense of the exterranean perhaps reappears in images such as those, which I call extraction landscapes, mobilized by the *Guardian*'s "Keep It in the Ground" campaign, which was in full swing as I began to work on this project. How we judge extraction is, in a sense, up to us. We can, with Book 2, see the stumps and sickly trees in Agricola's woodcuts as mere *natural signs* pointing to the presence of veins. Or we can think back to Book 1 and hear, instead, the concerns of Agricola's detractors, we can choose to read the woodcuts more in the way that Serres might argue, and we can, likewise, see the woodcuts in light of the campaign itself, such that the exterranean is not just a technical operation but something that forces humans and nonhumans into a difficult and never completely local engagement. Surely—and here we must be hypothetically historicist—Agricola's Ovid-quoting enemies might have looked at the woodcuts in the *De re metallica* not with the detached and descriptive eye of certain readers but as proof for their own claims about the destructiveness of extraction.

CHAPTER 5

DEMONIC MINES

This chapter turns from ripped-into mountainsides to the subterranean galleries in which miners, aided by all sorts of tools and machines, detach matter from the Earth. In particular, my focus here will be on the ways in which one specific early modern belief—the existence of so-called mining spirits or *daemones* in the underground world of mines—participates in fashioning a sense of the exterranean that allows, as in the Anthropocene, for an awareness *both* of the role of human agents as something like a geologic force *and* of the nonhuman responses to that activity. The texts studied here will, I think, testify to the fact that early modern miners, rather than subscribing to a discourse in which humans master Nature, considered the extraction of matter *ex terra* as extraction from a *terra* that is vibrant—a complex and potentially dangerous process involving confederated agencies. The belief in mining spirits was, in the early modern period, widespread, as is evidenced by Athanasius Kircher's (1602–1680) *Mundus subterraneus* (1665), which details how Kircher sent out a questionnaire to the Jesuit father Andreas Schaeffer in Slovakia, asking that he distribute it to various mine directors and miners. To question number 6—"Do you believe in the existence of little underground demons?"—each and every miner responded: "Yes."[1]

My goal here is not to produce a cultural history of demonic mines—such a cultural history would be possible, although vast, and would have to mention creatures as varied as tommyknockers, leprechauns, brownies, bluecaps, kobolds, hobgoblins, and many others.[2] Rather, via the writings of Agricola, whom we met in the last chapter, as well as through a text written by a late-sixteenth-century French mining expert, François Garrault, and via a brief discussion of Paracelsus, my aim will be to explore how such a belief is articulated in ways that force a reckoning with the vibrancy of matter in general and with that of toxic dust and vapors in particular. Jane Bennett, whose work on vibrant matter surely needs no further introduction, has argued that humanity and nonhumanity have "always performed an intricate dance with each other" and, further, that there was "never a time when human agency was anything other that an interfolding network of humanity and nonhumanity." I will clearly build on Bennett's work here—but whereas she notes that "today this mingling has become harder to ignore," I hope to show that (a) for early modern miners who frequently suffered major health problems because of the very real vibrancy of underground mines, such mingling was already hard to minimize, and that (b) the belief in underground *daemones* is one important symptom thereof.[3]

DEMONIC CREATURES IN THE *DE ANIMANTIBUS SUBTERRANEIS*

Several years before his *De re metallica* appeared in 1556, Georgius Agricola published a much shorter work titled *De animantibus subterraneis* (1549), or, in English, *On Subterranean Creatures* or *On Beings That Live Underground*.[4] It is, as one commentator put it, a book of "subterranean zoology,"[5] but it also tells us a lot about the perceived vibrancy of the subterranean spaces in which mining takes place and thus also about the reactivity of the *terra* in the exterranean. The book's opening lines would seem, on first reading and in opposition to the major claim of Jane Bennett's notion of vibrant matter, to produce what Jacques Rancière would call a *partition/distribution of the sensible* (*partage du sensible*), that is, to define a hard demarcating line between what is inert and what is living: Subterranean bodies, writes Agricola, can be classified into "the animate and inanimate" (*animatum [et] inanimatum*) and, further, what is "void of life" can be additionally "subdivided into that which spontaneously erupts from the earth under its own power [*quod sua sponte erumpit ex terra*], and that which is dug out of the earth [*quod ex eadem effoditur*]."[6] Having made this claim, Agricola then goes on to note that he has previously spoken about the inanimate categories of things in two previous works, a reference to, first, a work titled *De natura eorum quae effluent ex terra* (*On the Nature of Things That Escape from Earth*) (1546), which studies the properties of matter that escapes from Earth as if of its own volition, for example, gases, water, and the other materials they bring with them, including fire;[7] and, second, his *De natura fossilium* (*On the Nature of Fossils*) (1546), about things extracted by humans, the word *fossil* being derived from and still having the more general meaning of *fodio, fodere* (to dig).[8] Following Agricola's initial statement and self-referencing, there is thus *the inert* (already dealt with in earlier works) and then *the living* (to be dealt with in *De animantibus subterraneis*)—but, as we shall see, when it comes to the topic of mining spirits, indeed discussed in the 1549 work, the partitioning reveals itself to be less absolute.

To appreciate the place that mining demons occupy in the overall structure of the *De animantibus subterraneis*, it is worth noting that they appear only at the end of the work, a fact that highlights their importance. Agricola divides subterranean beings into three categories. There are, first, those that seek refuge underground only at night; second, those that go underground during certain seasons of the year; and finally those "that are really and truly called subterranean," that live underground on a permanent basis.[9] In the first two categories of subterranean beings we find a huge range of creatures: rabbits, foxes, beavers, badgers, otters, kingfishers, bats, horned owls, night ravens, screech and little owls, Alpine mice, Black Sea mice, shrew mice, domestic weasels, lemmings, porcupines, copper lizards, water lizards, chameleons, bank-dwelling swallows, as well as various kinds of habitual sea dwellers, such as the moray eel, the sea perch, the conger eel, the rock-dwelling wrasse, the sea wrasse, the tunny fish, the skate, the purple murex, the scallop, and the dolphin. In the final category—the *truly* subterranean animals—we find, as we might expect, moles, as well as beings such as the subterranean or wild mouse, as well as mussel-, round-, and earthworms.[10] So far, Agricola seems to remain faithful to his initial distinction between the inanimate (gases and other matters discussed in previous works) and the animate (the creatures discussed here)—but that division blurs in the final pages. At the end of *De animantibus subterraneis*, Agricola introduces a final kind of creature:

> Finally, amongst the subterranean animals or, as the theologians would say, amongst those beings that have souls, we can also count demons [*daemones*] that live in certain mines. Moreover, the class of these creatures is divided in two. For there are those that are aggressive and terrifying to look upon, which are generally hostile and unfriendly to miners. . . . Then there are peaceful [*mites*] demons, which are called *kobaloi* in Germany, and also in Greece, because they imitate humans. They laugh with unbridled joy and act as if busy with many things—but actually, do nothing at all.[11]

Discussion of these *daemones*, which fills up the book's final two paragraphs, sits as the work's endpoint. More than an afterthought or an excrescence, those paragraphs read like a point of arrival that invites connection with Agricola's writings on mining, for their emphasis is squarely placed not just on the *daemones* as subterranean creatures but on them as creatures who live in mines and on their interactions with miners. Via their quasi-material presence—in its very *quasi-ness*—they embody the material dangers that miners face as they extract matter *ex terra* in the mines. Moreover, while these demons can be hostile or peaceful, all can affect the bodily health of miners. Agricola gives a number of examples of specifically hostile demons, starting with one that "looked like a horse with a long neck and savage eyes" and that at a mine at Annaberg, in the Erzgebirge, "killed more than twelve laborers *with its breath* [*flatu*]" that it "emitted . . . from a gaping mouth [*ex rictu*]."[12] The demon's efficacy is very specifically not that of its body per se but of what exits from that body, the demon's breath (Latin *flatus*). The danger, here, is thus not just that the miner goes into Earth's entrails or merely that he meets a demon but that that demon, via its breath that the miner then breathes in, enters into the miner's lungs. Given that this breath brings death to the miner, it functions something like a kind of antipneuma, a "breath of death" rather than of life. But as with pneuma, the taking in of this breath "signifies being immerged in a place that penetrates [the breather] with the same intensity that [the breather] penetrates it," and thus the miner's presence in the mine is of the order of the pneumatological.[13] In other words, the demon's breath is a counter-movement to the miner's descent into the mine, that through which the mine enters the miner. A rationalizing explanation would conclude that this demonic breath that risks filling the mines and killing miners is (nothing but) a fantasy that recasts in fictional form the invisible but deadly toxic vapors and dirty air of the mines. An alternative theoretical move, following Latour in *Reassembling the Social*, would suspend this desire to rationalize and allow the demons (as *actants*) to compose the field in which they appear.[14] It is, moreover, not only these hostile spirits that await the miner.

> [There are also what some call] little mountain devils because of their short stature. . . . They appear as old men, dressed like miners, i.e., they wear a shirt and a leather apron that hangs down over their thighs. These demons are not in the habit of hurting. Rather, they wander down the wells and around mines and although they do nothing, they seem to train themselves in all manner of labor. Now they dig holes, now they pour what is dug out into vessels, and now they maneuver the hauling machine. Although sometimes the demons harass [*lacessunt*] the miners with extracted matter [*glareis*], only rarely however do they hurt [*laedunt*] them. The demon never harms the miners unless provoked by loud laughter or insults.[15]

These old men and primarily *good* demons thus also *lacessunt* (provoke, irritate, even attack) the miners not with breath but with *glarea*, a term that literally means gravel but that likely refers more specifically in this context to the matter broken off *ex terra*. The recent French translation of this passage goes so far as to render *glareis* as *grains d'or*, grains of gold.[16] Of what exactly this harassing consists is not explained explicitly, but the most likely situation, given how difficult it would be really to attack miners by *hitting* them with gravel or gold dust, is that the demons force the miners to inhale these small particles, which then irritate the lungs.[17] The extracted matter, thrown to be inhaled, thus functions in a manner similar to the demon's breath. In the *De animantibus subterraneis*, then, both hostile and seemingly friendly demons perform a reversal by which exterranean matter enters the miner who enters the mine, an embodied figuration of the chemical dangers that lurk within these vibrant mines.

DEMONS AND ASTHMA IN THE *DE RE METALLICA*

The *De animantibus subterraneis* ends by asserting both that mining devils reside primarily in underground spaces "from which metals are already being dug out or where there is hope that they can be dug out" and also that miners, thus encouraged and made "more eager" and "not frightened" by these presences, "work more vigorously" (*vehementius laborant*).[18] This book on *subterranean animals* thus concludes with the figure of the anthroturbator, the human subterranean animal, who works *vehementius* because of the presence of the demons—although constantly threatened, too, by their potential descent into his body. The human is not all-powerful, here, not a sovereign subject descending into an inert and deadened Earth but a human whose descent into the mines involves also this entanglement with demonic presences that appear allied with, even constituted of, the earthly matter they extract. When the *De re metallica* is published in 1556, it includes, between Book 12 and the extensive index provided as the end, the *De animantibus subterraneis*, such that both books share this same emphasis on the intermingling of human and demonic presences within the mines. Agricola also broaches the topic of mining demons in the main body of the *De re metallica*, more specifically at the end of Book 6, which it is worth examining here for the way in which it sets side by side topics that might at first seem unrelated but whose juxtaposition forges a sense of the exterranean as a confederation of heterogeneous agencies within mines recognized as vital.

The principal topic of Book 6 of the *De re metallica* is that of the tools (hammers, crowbars, picks, and buckets) and machines (especially for hauling and ventilation) used by miners—but that is not the only one. The book's closing pages deal with the ways in which miners descend into mines, then with various illnesses and dangers that miners might have to suffer and of which the final one to be mentioned is the presence of demons. As already in the *De animantibus subterraneis*, here the question of the miner's descent into the mines is juxtaposed with the descent of the mine into the miner, culminating in the figure of the demon. A woodcut (Figure 23) displays the four different ways in which miners might descend underground: (1) by using a ladder fixed to the mine shaft; (2) by being lowered on a rope attached to a stick or wicker basket; (3) when shafts are inclined, by "sitt[ing] in the first . . . and slid[ing] down in the same way that boys do in winter time"; (4) by walking down steps cut into the rock.[19] Both text and image, then, serve to show the miner entering the mine. There is no similar collaboration of text and image to describe the descent of the mine into the miner. Instead, there is only text, for example when Agricola describes how "the dust [that] is stirred and beaten up by digging penetrates into the windpipe and lungs, and produces difficulty in breathing, and the disease which the Greeks call ἄσθμα [asthma]," adding that "if the dust has corrosive qualities, it eats away the lungs, and implants consumption in the body."[20] A related danger is that of "air . . . infected with poison," brought about by the fact that "large and small veins and seams in the rocks exhale some subtle poison from the minerals," a sure danger, since "the bodies of living creatures who are infected with this poison generally swell immediately and lose all movement and feeling."[21] Book 6's mention of the demons is shorter than, but opens out onto, the one in the De animantibus subterraneis, to which Agricola directs readers eager for more detail. After explaining, concisely, that in "some of our mines . . . though in very few, there are other pernicious pests," that is "the demons of ferocious aspect [daemones . . . aspectu truci]," Agricola explains: "I have spoken [about these demons] in my book De animantibus subterraneis."[22] The only extra information he provides at this point is that "demons of this kind are expelled and put to flight by prayer and fasting [precibus et ieiunis]."[23]

Agricola never says explicitly that the mining spirits embody chemical fears—by all accounts, he simply believed in them. Their juxtaposition, however, speaks volumes, capturing the very real sense of danger involved with extracting matter ex terra, specifically that terra not be inert, that it enter into the miner and cause pulmonary problems. The juxtaposition of bodily illness and demonic presences is replayed—and intensified—in a final section of Book 6 that focuses on the reasons "why pits are occasionally abandoned."[24] To the most obvious reason—that the mines no longer "yield metal"—Agricola adds six more: (1) that the mines have become flooded; (2) because of "noxious air"; (3) the presence of "poison produced in particular places"; (4) "the fierce and murderous demons [daemon truculentus et homicida]—for if they cannot be expelled, no one escapes from them"; (5) the "underpinnings become loosened and collapse"; and (6) "military operations."[25] Structurally, floods, noxious air, poison, demons, collapses, and military activity are comparable members of a list of dangers, but the demons also appear here to emerge as presences that stand in for and translate the miners' fear regarding the more obviously material dangers. They figure and concentrate this variety of dangers in just one shape.

FIGURE 23. Ways of descending into the mines, from Agricola, *De re metallica* (1556), 171. With kind permission of the Rare Book and Manuscripts Library, Columbia University.

Agricola was far from the only humanist to express belief in mining demons and to secure for them a key role in exterranean ecologies of extraction. With the strategy of confirming the rule by citing the exception, I now turn to explore a case where an early modern text specifically denies the presence of mining spirits in order to demythologize, deaden, and assert human mastery over planetary matter. Let us shift our attention to France and to Charles IX's sponsorship of France's mining industry in the 1570s. By this point in time there were as least ten mines in the Lyonnais, including at least one gold mine and six silver mines.[26] At the heart of this boom is a certain François Garrault, who was, among other things, the *trésorier de l'épargne* and controller general at the Cour des monnaies, a sovereign court that had been set up in 1552 to regulate the minting of coins. In his official roles, and in a number of published works, Garrault was most frequently interested in economic questions concerning the value and circulation of coins, as in his *Paradoxe sur le faict des monnoyes (Paradox on the Question of Coinage)*, which deals with topics such as the relative valuation of different coins under Henri II.[27] In other words, Garrault was mainly interested in how a royal court assigned in the name of the king economic value to seemingly inert objects; that is, he was concerned with the relationship between abstracts *unités de compte* (for example, *livres* and *sous*) and *unités de règlement* (the actual coins, for example, the *écu*). But, as we shall see in due course, his writings also tell us a lot about the exterranean character of those *unités de règlement*.

Before exploring what Garrault says about mining spirits, it is important to recall that, for various reasons, the balancing act linking economic value to exterranean objects had become particularly difficult in the latter part of the sixteenth century. In 1578, at the height of the Wars of Religion and only six years after the Saint Bartholomew's Day Massacre, during which something like ten thousand Protestants were killed in the streets of Paris, Garrault wrote that "one of the greatest evils that these civil wars have brought to our kingdom, is the disorder [*dérèglement*] of our coinage, [which is] the foundation of our political state."[28] Lest his readers not understand the immediate significance of such questions, Garrault gives an example that reads somewhat like a child's math problem: If, in the month of June 1575, when the *écu* (a specific type of coin) is worth 60 *sous*, Peter lends to John the sum of 60 *livres* in the form of twenty gold *écus*, and then in 1577, the *écu* has a new value of 66 *sous* and John wants to give back to Peter the 60 *livres* he borrowed in the form of 18 *écus* and 2/11ths, then while this indeed adds up to 60 *livres*, Peter has nevertheless been cheated.[29] Given that Garrault spent much time and ink on such questions, one might assume that he is generally attentive *not* to things, *not* to matter, *not* to the planetary origins of silver or gold, *not* to the fact that coins are made of matter extracted from the Earth, but that would be to ignore one of his most original—and for our purposes, most interesting—pieces of writing.

Garrault published his *Des mines d'argent trouvées en France, ouvrage et police d'icelle* (*On the Silver Mines Found in France, and on Their Functioning and Management*) in 1579 as an apology of that particular industry and to sing in particular its excellence in France. Although Garrault nowhere mentions Agricola's name, the opening pages of the text rehearse, in terms very similar to those of the *De re metallica*, the question of the "commodity or incommodity of Metals."[30] While some might argue (for example, Ovid and those who quote him) that metals lead to "murder, envy, theft, and all other kinds of evil" (Ovid's iron age), Garrault asserts that they are "more necessary than [they are] damaging [*dommageables*]."[31] Like Agricola, Garrault defends his position by stating that Nature "with accustomed goodness" nourishes all humans and that She will provide the resources they need to survive within their particular geographical locale.[32] The details are important: Fertile land and an abundance of fruits are given to the inhabitants of plains and flatlands, pastureland and livestock are offered to valley dwellers, but only "dry and sterile" land is dispensed to those who "retreat into the mountains." Yet, mountain dwellers are not so short-changed as first appears. Their land may be unfit for agriculture, but Nature makes sure that the mountains are filled with "mines both rocky and metallic [*tant pierreuses que metaliques*]" that can become sites of extraction. Nature's control of geology and geography ensures that the *montagnards* can extract and then sell precious metals, in order to purchase those resources that they would otherwise lack ("moyen d'achepter par icelles ce qu'il leur default").[33] Garrault's "Nature," a close affiliate to the Terra Mater seen in the first section of this book, is active and generous by providing metal deposits *in a dull planet* for humans. Mountain metal must be mined, argues Garrault, because it is abundant and because the extraction process is easy. He adds that Nature not only fills the mountain with extractable matter but also that She composes a landscape defined by other matter useful in the extraction process. Nature thus causes trees to grow on mountains despite the dry land, for the provision of wood necessary for supporting the hollowed-out shafts and tunnels ("pour estamper les creux"). That wood is also essential for stoking the fires of the refining process. Nature also provides streams, necessary for the grinding mills ("pour dresser & edifier martinetz"). Even the slopes of the mountains are shaped to allow excess water to drain away from the mines. Nature, here, shaped the planet to provide for man. Geology and geography not only serve humanity but were fashioned by Nature for precisely that purpose.

Nature here functions as a fantasy that says: *Extract! Nature Put the Precious Metals There for You!* This is the early modern equivalent of the 2008 Republican slogan *Drill Baby Drill!*

It is within such a definition of "Nature" that Garrault approaches the question of demonic mining spirits. To understand this, we must realize that Garrault passes from the global perspective on what is commodious or not for humans in general (after Ovid) and from the national perspective (the mines found *in France*, as the title signals) to focus in particular on the town of Chitry, now called Chitry-les-Mines, situated about thirty miles northeast of Nevers. From 1493 until the early seventeenth century, Chitry was by all accounts one of the most important centers of French mining.[34] Its importance grew upon receiving a royal privilege for silver extraction on August 26, 1579, which led to something of a silver rush. According to mining historians, Chitry proved to be "a successful venture," producing approximately 1,100 *marcs* of fine silver and 100,000 *marcs* of lead per annum.[35] What the miners at Chitry searched for primarily was, more specifically, "lead glance," or galena (French *galène*), the most important form of lead ore from which, once mined, silver can be extracted. Extraction of galena was not new in France—Chitry in a sense took over from another French site of extraction that operated throughout the tenth through fourteenth centuries, the silver mines in the Fournel valley in the Alps, which have now been turned into a tourist attraction centered around a town renamed L'Argentière. Lead, as I likely do not need to say, is a toxic element: While connected to the mineral's crystal structure, it is basically safe, but as the galena is extracted and refined, pulverized lead dust can be inhaled and cause major health problems.[36] At Chitry, as elsewhere, mining thus involved various dangers.

It is in this specific context—the question of the dangers of mining (as already in Agricola)—that the question of mining spirits comes up. Nature may well bend geography to "suit" humanity, but it also, seemingly, rebels. Spanish miners—and especially German ones, who helped the Spanish on many occasions—have dealt, says Garrault, with many of the problems that miners face, including unplanned fires, lung diseases, and so forth. But several problems remain unresolved. (1) Bad vapors issue forth from the Earth: "[Miners] have been unable to stop the bad vapors from issuing from their mines," resulting in numerous deaths;[37] (2) wild animal bites; and (3), which seems to synthesize (1) and (2), are those "metallic spirits" that infiltrate the mines— they are protean, taking on the shapes of, among other things, small-necked proud-gazing horses that kill miners with their breath and neighing. Other metallic spirits disguise themselves as mine workers dressed in black frocks who then proceed to "drop" miners down into the deep pits.[38] And then there are the *follets*, which Cotgrave renders as "Hobgoblins, Robin-good-fellows, and Bugbeares."[39] Garrault, the good comparative philologist, reminds us—in a manner similar to Agricola's—that the Greeks called these spirits *kobaloi*. Garrault's geographies of mining thus put space at the service of humans but also have these mischievous spirits wandering about animating that geography, potentially protecting it from humans.

The *kobaloi* as Garrault paints them are neither wholly evil nor wholly good. Thought of as spirits of the mines and as expert miners in their own right, the *kobaloi* can both wreak havoc and provide protection. As Garrault says, "they come and go through the mines, they go up and down, and they pretend to be working" ("ils vont & viennent par la mine[;] ils montent & descendent du hault en bas, & font toute contenance de trauailler").[40] As long as the miners do not irritate them, the spirits in fact "take care of them" ("ont soing d'eux"), but they can also end up "in a mad and grumpy mood" ("une humeure mauvaise et grossiere").[41] Such spirits could thus interfere in all sorts of ways: They might cause mine shafts to cave in; they might trick miners into extracting worthless ore that, in smelting, would release dangerous toxins and lung problems but no gold or silver. These geographically situated spirits can be tamed and their evildoing overcome via "prayers, fasting, and orations" ("on peult surmonter leur malice par prieres, ieunes & oraisons").[42]

Most interesting of all to me is that Garrault evokes the demonic mining spirits only to dismiss them. In France in general and in Chitry in particular, there is in fact no need to worry about any of this—France's geography and geology, he tells us, are such that silver mining, or so Garrault claims, is completely safe: "France's silver mines are not in any way dangerous" ("[les] mines d'argent de France ne sont aucunement dangereuses"), or, at least, all dangers are purely technical: Water can build up in the mines, and earth might fall if the mines are not properly constructed, but such problems are avoidable. Apart from these technical problems, argues Garrault, there is nothing to fear. France's geography and geology allow no place for the bad vapors and dangerous beasts ("Il n'y a aucunes mauuaises vapeurs, ne bestes dangereuses"), which means that the French are more than happy to mine ("qui faict que les habitants des lieux entreprennent volontairemēt l'ouurage").[43] The ore is crushed, washed, and dried before being smelted, with the goal of getting rid off ("euaporer") all that it contains that is "bad and dirty [mauuais, & infect]," namely arsenic, sulfur, and antimony ("comme arcenic, soulfre, & antimoine").[44] To promote the geology and geography of Chitry as a site suited to the extraction and smelting of silver, Garrault, who has clearly read and digested Agricola's *De re metallica*, demythologizes and deanimates the Chitry geography. He reasserts the absolute difference between what he sees as inert matter and what might animate it, reside in it, coexist in and with the galena dust. *No Kobaloi Live Here!*

DIE BERGSUCHT AND DEMONS *IN DER ERDEN*

The lines of thought explored so far in this chapter, via the writings of Agricola and Garrault, find their *point de fuite* in a book published by the physician, botanist, and alchemist Theophrastus Bombastus von Hohenheim, known as Paracelsus (1493–1541), namely the first monograph about the occupational diseases of miners, *Von der Bergsucht und anderen Bergkrankheiten* (*On the Miners' Sickness and Other Miners' Diseases*), written circa 1533–1534 but published only posthumously in 1567 by Samuel Architectus. As such, it was written before but only went into print after Agricola's *De re metallica* and Garrault's *Des mines d'argent*. To summarize rapidly, we can note that Book 1 deals with the pulmonary problems that affect miners (who work underground), that Book 2 treats the related question of those illnesses that attack smelter workers (who work above ground), and finally that Book 3 takes up the specific question of diseases caused by mercury. A full study of the multiple ways in which either the thought of Paracelsus in particular or early modern alchemy in general acknowledged and sought to forge a lexicon to account for the vibrancy of matter would take us too far from this chapter's specific context. It is nonetheless worth opening up *Von der Bergsucht* for a brief glimpse of how the work, while focusing mainly on the very material, mainly chemical, connections that link individual humans, as microcosms, to the macrocosm of Nature also grants agency to mining spirits.

Book 1 describes the miners' disease (*Berghsucht*) as the subterranean equivalent of the more general category of lung sickness. Whereas the latter affects those who live above ground, the former targets "those who are in the Earth [*in der Erden*]."[45] In both cases the illness is transmitted to the lungs via the air, which Paracelsus calls "chaos," an idea explained via a useful comparison: Just as "food is digested in the stomach and has its special gullet, thus the chaos is digested in the lung."[46] Disease arises because of what the air carries within it, which Paracelsus divides into mercury, salt, and sulfur—not to be confused with the modern matters having the same name—and which, via the air, can enter into the lungs. Via another material comparison, Paracelsus explains this process by discussing how with clear wine in a clean barrel, at the end of the year when the wine is poured out, "mercury, sulfur, and salt are found to have settled on this barrel," a substance he calls "winestone" and "tartarus": "In the same manner, just as there is something in the wine which was not seen in it, there is also a body in the chaos, which attaches itself to the lungs, as to its barrel, and which then hardens there like a mucus in its viscosity, after which the coagulation starts, which is the matter of the lung sickness."[47] As regards the air or chaos in the mines, it is said to become "a soup" of those minerals of which the hillside is made and which are being extracted.[48] In other words, the souplike chaos into which the miner descends to extract matter is itself exterranean.

According to *Von der Bergsucht*, the miner is a human who descends to work "in der Erden" and who is thus forced to nourish his lungs "with the chaos that is there," running the risk of bringing into his lungs "that which has been cooked in the chaos," that is, "the mineral impression," which Paracelsus calls the "tartarus of the lung" and which leads to the miners' disease.[49] Terminology aside, such an explanation tessellates in its broad lines with a twenty-first century understanding of occupational lung diseases, such as occupational asthma or silicosis. Both, eschewing early modern humoral explanations, focus on the material transmission of matter into the lungs. For Paracelsus, it is about exterranean mineral matter, dissolved into the air soup, that the miner then breathes; for modern medicine, to give one example, silicosis results from the inhalation of crystalline silica dust. My main point is not to situate Paracelsus as particularly modern but rather to emphasize that his discussion of *Bergsucht* is, up until this point, very specifically focused on matter. Material details are added chapter by chapter, including about symptoms (coughing, gasping, short-windedness, hoarseness, foul odor in the mouth, diarrhea, etc.).[50] However, and as Charles Webster has shown, health and disease for Paracelsus are intimately connected not only materially but also with the more universal battle between good and evil, a view in which living matter is constantly at risk: "Nature strives against nature" ("die natur wider natur strebt"), and "one thing in nature opposes another" ("eins in der natur wider das ander ist").[51]

The Fourth Tractate, which turns from the question of causes and symptoms to that of "the cure of the miners' sickness and its necessary parts," places emphasis on the need for practical knowledge. Already at the start of Book 1, Paracelsus had spoken of his weariness regarding established authorities: "The birth of the miners' disease is a testimony of the errors of the Ancients in the writings on the lungs."[52] But here, in the first chapter of the Fourth Tractate, Paracelsus goes further to suggest that practical knowledge can complement bookish thought: "In order to consider the welfare of the miners so that they can be protected from the aforementioned miners' sickness," it is not enough "to speak with the learned men and the philosophers, but with experienced men."[53] This announcement seemingly paves the way for the very sudden appearance, in that tractate's second chapter, of mining spirits. The second chapter takes up the topic of prevention, detailing a total of five specific measures, a list that begins as follows: "First, the mine spirits must be forestalled so that they leave the miner uninfected; this is begun before they become subject to the mine and the ore spirits."[54] The second through fifth preventative measures are much more material, relating to body parts, diet, medications, and arcana. Paracelsus's *Von der Bergsucht* thus provides confirmation of the motivated nature of the juxtaposition of medical questions and demonic hauntings.

CONCLUSIONS

Agricola may well assert at the start of his *De animantibus subterraneis* that a solid demarcation can be made between the *animatum* and the *inanimatum*, but his and Garrault's and Paracelsus's discussions of demonic mining spirits in close proximity to reflections upon the various physical and specifically chemical dangers that await the miner tend rather to blur that distinction. The theoretically *inanimatum* (the matter of which the mines are hewn, the exterranean matters extracted) reveals itself to be *animatum* via this demonic haunting, an uncertain vibrant materiality that is figured by the demonic spirits. In the introduction to her *Vibrant Matter*, Bennett describes the hunch at the origin of her project, namely that "the image of dead or thoroughly instrumentalized matter feeds human hubris and our earth-destroying fantasies of conquest and consumption," which she sees as "an impediment" to "more ecological . . . modes of production and consumption."[55] While the miners' fear regarding the possible dangers of mining and the demonic spirits in which they believed can hardly be said to betoken, directly, an ecological mode of production, they do—with what Bennett would call their "taint of superstition"—give a voice (or rather, a face) to the vibrancy of the poisonous vapors and other physical dangers that make it impossible to think of matter as inert.[56] This chapter, in its close reading of Agricola, Garrault, and Paracelsus, has attempted to develop an attentiveness to the liveliness of those mining spirits and to suggest that they contribute to a sense of the exterranean *not* as one in which humans fully master an inert Earth but as one in which taking matter *ex terra* involves, at the site of extraction, an Earth that is vital, vibrant, and ready to enter the miner as much as the miner enters it.

PART III

HIDING IN EXTERRANEAN MATTER

We *anthropoi* are not only Earth-bound.[1] As we scuttle around in our daily com-posting, we pick up, carry about, put down, cut up, reshape, scrape, feel, look at, sit on, build with, bury, inscribe, send currents through, touch, lick, eat, taste, ignore, throw, and otherwise inter- or intra-act with countless chunks of Earth.[2] We are Earth-bound; we are also bound-to-bits-taken-from-Earth as we—constantly—interact and sympathize with endless exterranean things.[3] The first two sections of this book were named after and framed by contem-porary French theater. This third section finds its emblem and impetus else-where, in a photographic image that can here serve commodiously to question perceptual practices. Liu Bolin's *Hiding in the City n° 95* shows a man (in fact, Liu Bolin himself) stained by yet barely visible in front of a pile of extracted and combustible black sedimentary rock (coal)—he is as if wholly wrapped in exterranean matter to the point of being almost indistinguishable from it. The image is part of a series titled *Hiding in the City* that Bolin began in 2005 in response to the Chinese government's demolition of the artist village Suo Jia Cun. Begun in that localized context, Bolin's series expanded to tackle more generally the interlocking of camouflage and discernment, to represent the (in)visibility of the human in relation to the matter and spaces in/with which s/he/it lives.[4] A useful way to imagine the two chapters that follow is thus as something of a scholar's counterpart to Bolin's image, as an attempt to make perceptible some of the ways in which, as humans, we live with and in exterranean matters. One thread through what follows connects points at which the *ex* of exterranean oscillates between connecting and disconnect-ing, between *ex* as *taken from* (which maintains a material genealogy despite a break) and *ex* as *former* (in which a break interrupts perception of material continuity), to imagine, theorize, and act out how our *living with* all manner of exterranean matter sometimes perceives and sometimes does not perceive whence that matter came. A second thread, from the same yarn and knotted to the first, associates these moments of perception and asks about the kinds of collaboration between humans and exterranean matter that they reveal. More specifically, Chapter 6 focuses on limestone and on the way in which, as geomedium, this extracted matter becomes a support whose exterraneanity allows and becomes visible through its alternate durability and fragility. We will step close to and draw inspiration from Jeffrey Jerome Cohen's study of stone and Tiffany Werth's work on the lithic, but emphasis is shifted squarely onto the exterranean.[5] Chapter 7 takes up another exterranean matter, salt. Analyzing a number of different kinds of texts, including poems by André Mage de Fiefmelin and Conrad Celtis, a novel by Rabelais, and a saltcellar (*saliera*) made by Cellini, the chapter asks how salt is described via its chemical doings and to what extent and how its exterraneanity is (or is not) perceived.

CHAPTER 6

GEOMEDIA

The Anthropocene, as it invites us to reconcile human history and geological time, makes memory a function—pressingly—of media and of matter.[1] Forced to reconnect our insignificant human timelines with the expanses of deep time, we look back with sympathy at the dinosaurs and gaze forward, forlorn or indifferent, to extinction and to Man's "fossil futures." We are also more aware than ever of the speed with which media can disappear—as in Bruce Sterling's Dead Media Project.[2] Existing as a geological force, we realize anew the materiality of media and of memory: Our planetary footprint increases daily as e-waste proliferates, yet we live on the edge, as Vint Cerf, Google's vice president phrased it, of an "information black hole" in what might be a "forgotten century," giving rise to calls for "digital vellum" to ward off "bit rot."[3] Within this current context, slow and durable media, especially rock, that exterranean stalwart, possess a particular and graspable power, which draws us toward it. With Tiffany Werth, this chapter asserts that "[as] humans, we hitch ourselves to the endurance of stone, from which we gain the deep time of lithic memory."[4] It also asserts, with Jeffrey Cohen, that in stone's "aeonic endurance we discern something ardently desired, something ours only through alliance."[5] Like hoarders reassured in times of trauma or mourning by the accumulation of "things," as we affect the planet on a geological scale and wonder what future species will make of us, we are solicited to keep in mind that the matter in which our actions and deeds are or might be hewn is exterranean.[6]

CAEN'S CROCODILES

Although—or because—early modern humanists did not yet know of geology or of deep time the same way that we do, they have much to teach us about

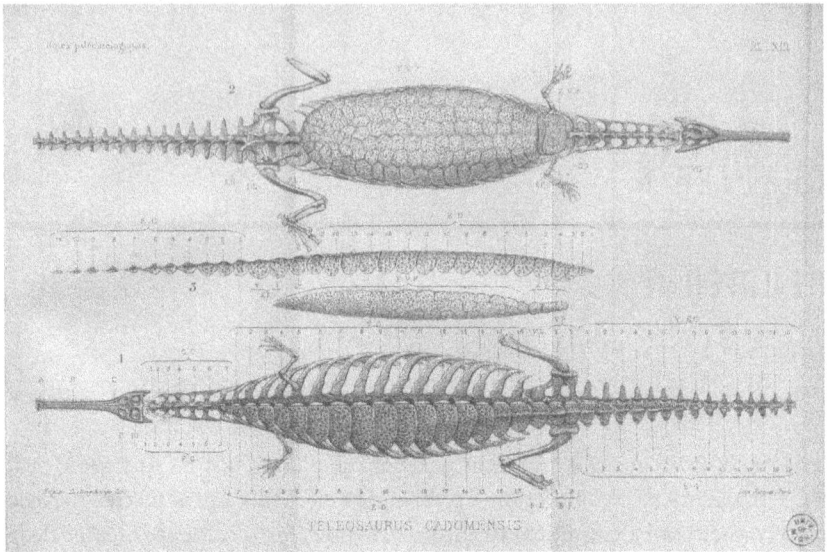

FIGURE 24. Drawing of fossilized crocodile found in Caen in 1817, from Eugène Eudes-Deslongchamps, *Notes paléontologiques* (Caen: Le Blanc-Hardel and Paris: Savy, 1863–1869), vol. 1, plate XII. © University of Michigan Library.

how to relate to lithic memory in general and to the exterranean materiality of media in particular. In order to explore one pre- or proto-Anthropocene moment when human history's relationship to exterranean media comes into particular focus, I turn my attention to the particularly petrous town of Caen in Normandy. To make sense of this turn, we must first acknowledge some chronological crossroads. On the one hand, we who live after the invention of modern geological sciences can articulate with scientific bravado as follows: Caen's (human) history is rockbound because about 165 million years ago, long before Normandy was synonymous with apple trees and cider, it was home to mangroves and crocodiles. We can say that warm seas used to wash limestone-rich mud onto the shores, which slowly deposited matter that would eventually be extracted, beginning in antiquity, from mines in and around Caen, in the form of a creamy-yellow limestone. We can say that Caen's geology, its rich ooltic veins, determined the town's architectural landscape, its particular yellow-white color, and beyond a fair chunk of European history.[7] We can assert this connection between deep time, geology, and human-scale history, between mangroves and quarries, and between the Jurassic and the early modern/modern because in 1817 a fossilized crocodile skeleton was discovered in a block of recently mined Caen stone (Figure 24).[8]

On the other hand, the inhabitants of sixteenth-century Caen, including celebrated Norman poets such as Jean Marot, Jean Vauquelin de la Fresnaye, and François de Malherbe, would not have known that Normandy was once tropical or that the rest of what would become France had once been under water.[9] The argument, in what follows, is distinctly *not* that Caen's humanists *knew* of deep time or geology or about crocodiles and mangroves. It is rather (1) that early modern Caen's architecturescape, built largely from local limestone, already made Caen a petroglyph whose exterranean durability inhabitants perceived and indeed counted on and (2) that at least one Caennnais humanist already intuited that the Earth and matters extracted from it function as recording media—and indeed as particularly fragile ones.[10] In other words, we shall see that Norman humanists already intuit that—as Chakrabarty will articulate for the Anthropocene—human time and deep time are inherently interwoven and, further, that this interweaving partakes of an exterranean sensibility. Like the coal-covered figure in Liu Bolin's *Hiding in the City n° 95* reproduced on the cover of this book, early modern Normans arguably walked around the city of Caen with an acute awareness of their entanglement in/with exterranean matter. As such, they can offer us valuable reading lessons, as this chapter sets out to demonstrate.

A map from the late sixteenth century is a good place to start (Figure 25). It captures and displays, in multiple ways, a keen cognizance of the fact that the town is exterranean, the result of Normandy rock going from below to above ground, as if Caen were created by turning Normandy's surface depths upside down. The map depicts, chorographically, numerous buildings made of stone extracted locally, generally from underground galleries *within the city limits*, where veins were frequently ninety to one hundred feet thick.[11] Prominent are the huge limestone walls and towers of the castle built by William the Conqueror circa 1060, one of the largest castles in Western Europe, which still dominates the town today. Among many other visible structures are the Eglise Saint-Pierre de Caen, opposite the castle, the Abbaye-aux-hommes, and the Abbaye-aux-dames. We see buildings made of local stone—not just humanmade structures but, if we simplify our gaze, exterranean objects that sit on the planet's surface, both *of* it and *on* it. The map goes beyond chorography, however. It also specifically draws our attention to the fact that Caen is its own site of extraction. The legend on the left-hand side of the map identifies Caen's castle as follows: "The Castle, seated upon stone quarries" ("Le Chasteau, assis sur carieres"), an inscription that foregrounds and replays the vertical shift of stone from below ground to surface level that constituted the castle, an intuition also graspable by modern visitors to the castle, who can see exposed (yet not-extracted) stone in a moat wall directly opposite the castle (Figure 26).

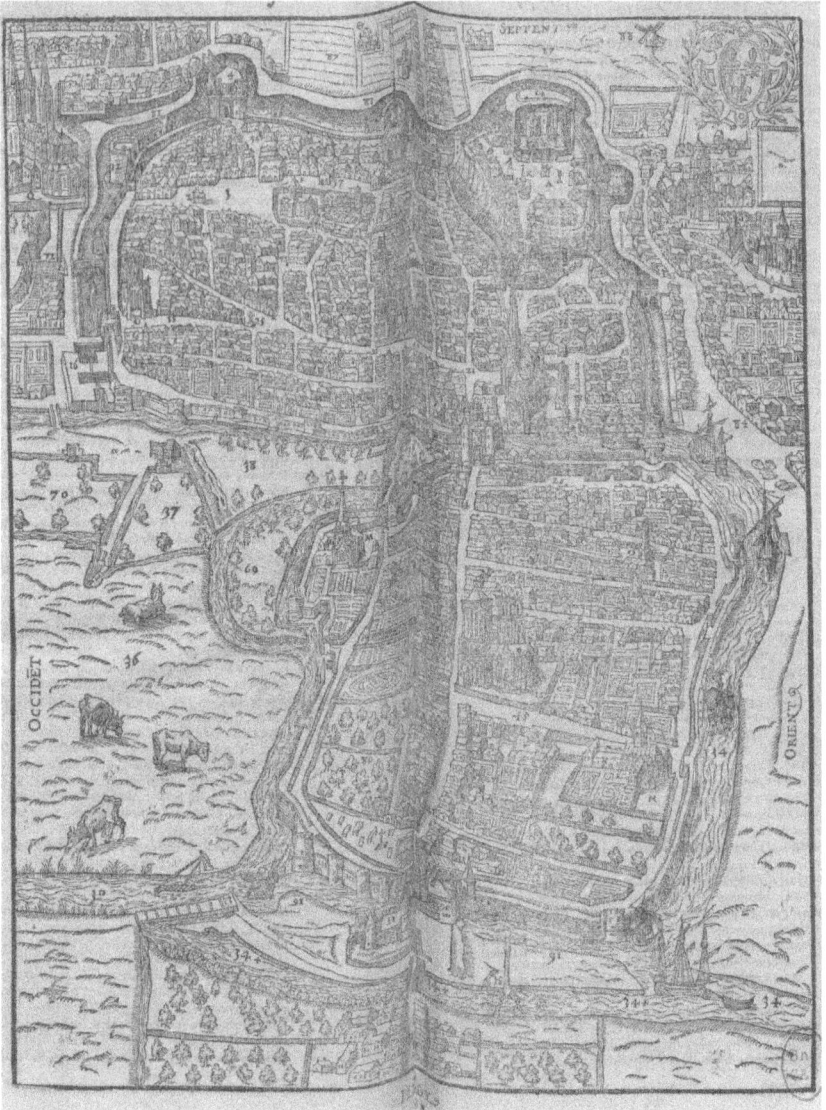

FIGURE 25. Map of Caen, from François de Belleforest, *La cosmographie universelle* (1575). Source: Gallica / Bibliothèque nationale de France.

FIGURE 26. Exposed stone in a dry moat at Caen Castle, June 2015. © Phillip John Usher.

Similarly, item number 25 in the legend points to the fact that Caen was not just made of stone, not just a site of stone extraction, but a site for the commercial exchange of stone: "The Beautiful Cross [which marks the] Limestone and Tile Market" ("La belle Croix, & marché a la chaux, & tuile"). Caen, then, was and is a form of geomedia, a medium made from the Earth (geo < Gk. *Gaia*) because made from exterranean matter.

Early modern Caen is a particularly apt site for thinking about geomedia and for reflecting on the reciprocity of human history and the mineralogical because this petroglyph of a town experienced two major stonecentric events in the early modern period, at the heart of which we find humanists in various guises. The first of these events, which is normally related in contexts and timelines wholly human rather than earthly, concerns the sudden proliferation of building projects in sixteenth-century Caen. Between approximately 1500 and 1560, numerous new buildings were built out of locally sourced stone, changing the city's appearance *by bringing more of Normandy's underground above ground*. The second event can be dated to various points in 1562: Protestants, driven by an iconoclastic fury that marked the start of the French Wars of Religion, destroyed as many and as much of the town's buildings as they possibly could, all of a sudden making Caen's stoniness seem more fragile and less enduring, causing a sudden rethinking of the extent to which geology and the planet could function as durable recording media. Human sympathy with the lithic, which had been taken for granted for centuries, was—as now in the Anthropocene—suddenly shaken up, more easily perceivable because of its endangeredness. To follow up on these two events and the shifts in awareness that they betoken, with the goal of putting received humancentric histories of them into conversation with the exterranean-as-medium and -as-memory, it will be necessary to consult both buildings and texts.

As goes without saying, architecture was central to the intellectual and aesthetic changes that marked early modern Europe. I will not attempt a concise history thereof here, but we can recall that Vitruvius's first-century *De architectura* was rediscovered, printed, and translated and that its emphasis on orders, symmetry, and proportion quickly spread throughout Europe, influencing the construction and appearance of buildings not only in capital cities but also— and sometimes first—in the provinces. Humanist-trained architects gradually replaced masons, and building knowledge, which used to be passed on orally and manually from generation to generation (normally, of course, from father to son or father to young male apprentice), now circulated in book form.[12] To embody this moment of change in one iconic structure, necessarily oversimplifying, we might bring to mind how the primarily defensive medieval Louvre morphed into something of a Renaissance palazzo when François Ier had

FIGURE 27. Hôtel de Nollent, Caen, in 2015. © Phillip John Usher.

its dungeon knocked down and Henri II subsequently built a new Italianate *aile* whose outward-publishing façades were more important than their ability to ward off attackers and whose style directly related to the new books being published about architecture.[13] Anyone familiar with the Renaissance Louvre or with the numerous palaces that French kings built or remodeled in the Loire valley at this time will recognize that early modern Caen—as were many other towns in France's provinces—was shaped by similar influences.

Throughout the first decades of the sixteenth century, Caen witnessed the construction of several buildings where earlier gothic and newer Italianate elements collide, as in the Hôtel de Than, built for Thomas Morel, finished circa 1527. The gothic asymmetrical façade and heavy square tower stand in aesthetic opposition to the many Italianate details, such as the medallions, nods to triumphal architecture, the foliage friezes, and putti. Also emblematic of this moment is the Hôtel de Nollent, otherwise known as the Manoir des Gens-d'Armes (Figure 27), built by Gérard (or Girard) de Nollent at the beginning of the reign of François Ier.[14] Again, the somewhat gothic structure is marked by the architectural trends of its moment of construction, for example in a series of medallions that allude directly to the first French printed edition of Petrarch's *Trionfi*.[15] Traditional histories emphasize, then, the spread of

FIGURE 28. Petrarchan *triumph* engraved on Norman limestone at the Hôtel de Nollent. © Pippa Grantham.

classical and Italian architecture from Italy, through France's capital and Loire valley, to places like Caen. What retains my attention is that this human-scale flow—aesthetic, economic, and cultural in a broad sense (*translatio imperii et studii*)—*arrives* and is *petrified* in stone extracted in Caen and neighboring towns. Geology was, in other words, a key player in Caen's early modern architectural renewal.

To appreciate the interplay of the human and the geological, a brief exercise in reading is worthwhile. As already noted, around the Italianate medallions atop the Hôtel de Nollent we find inscriptions from Petrarch's *Triumphs* (*Trionfi*), a fourteenth-century poem recently published in France and in which Petrarch witnesses a number of victorious pageants: Love triumphs, then Chastity triumphs over Love, Death over Chastity, Fame over Death, Time over Fame, and Eternity over Time, a series that critics have read as both a narrative about the different states through which Man passes and as a more personal one about the ups and downs of Petrarch's relationship with Laura. Take a look at Figure 28, which shows the third triumph (*Mors vincit pudiciciam*) engraved on Normandy limestone at the Hôtel de Nollent. What could be more "Renaissance" than this—Petrarch engraved on an Italianate structure, in France? We can of course—and indeed should—tell the story this way. But when we look at the medallion, we should also remember that we are looking, after all, at a piece of sedimentary rock, composed largely of calcium carbonate ($CaCO_3$). We are more used to seeing the *factum*, the human inscription upon that rock, than the exterranean matter that is the support for that inscription.

Seeing both the limestone and the Petrarchan *factum* at the same time—a critical intervention—puts us precisely at the theoretical vantage point offered by the Anthropocene, which involves a redistribution of the natural and the social/symbolic/cultural. In our present moment, this redistribution is exemplified by the appearance of a new kind of "stone." Scientists have baptized this new "stone" "plastiglomerate," to capture the fact that it is "an indurated, multi-composite material made hard by agglutination of rock and molten plastic" (Figure 29).[16] Such stones are particularly common on the beaches of Hawaii—most well known is Kamilo Beach on the southeastern coast—where waste plastic is washed ashore from the (so-called) Great Pacific Garbage Patch.[17] One report says that in a single bag of Hawaiian sand, 90 percent proved to be plastic.[18] As Bruno Latour has put it, with such a "stone" it becomes quite simply impossible to "distinguish [*départager*] man [the human] from nature."[19] Only seeing the triumph *on* the limestone and not the limestone, as the founders of Renaissance Studies would have wished—both Burckhardt and Michelet saw in the Renaissance the moment Man or the human is separated from Nature—we see only the human imprint, not the human entanglement with the exterranean. To *read* the stone in such a fashion is to subscribe, perhaps unwittingly, to Michelet's idea, presented at the opening of his *Introduction à l'histoire universelle* (*Introduction to Universal History*) (1831), which states: "As the world came into existence, so did a war that will only end with the world, and not before: that of Man against Nature [and] of Spirit against Matter."[20] Nature is on one side of the divide, Culture on the other, as Michelet captures with the example of mountains: While "the Alps have not grown any bigger," humans reshaped them by creating "the Simplon Pass" (the pass, and later tunnel, between the Pennine Alps and the Lepontine Alps in Switzerland).[21] To see *only* the triumph is to read the way that Michelet read the human creation of the Simplon Pass as a human imprint on an indifferent, passive, and unchanging Nature.[22] As much as Burckhardt and Michelet have been challenged on various points, their shadow still looms incredibly large over early modern studies, perhaps to the point that we frequently do not even see it. Despite the growing amount of critical attention being paid to nonhuman animals in early modern studies and indeed to nonhuman life forms more

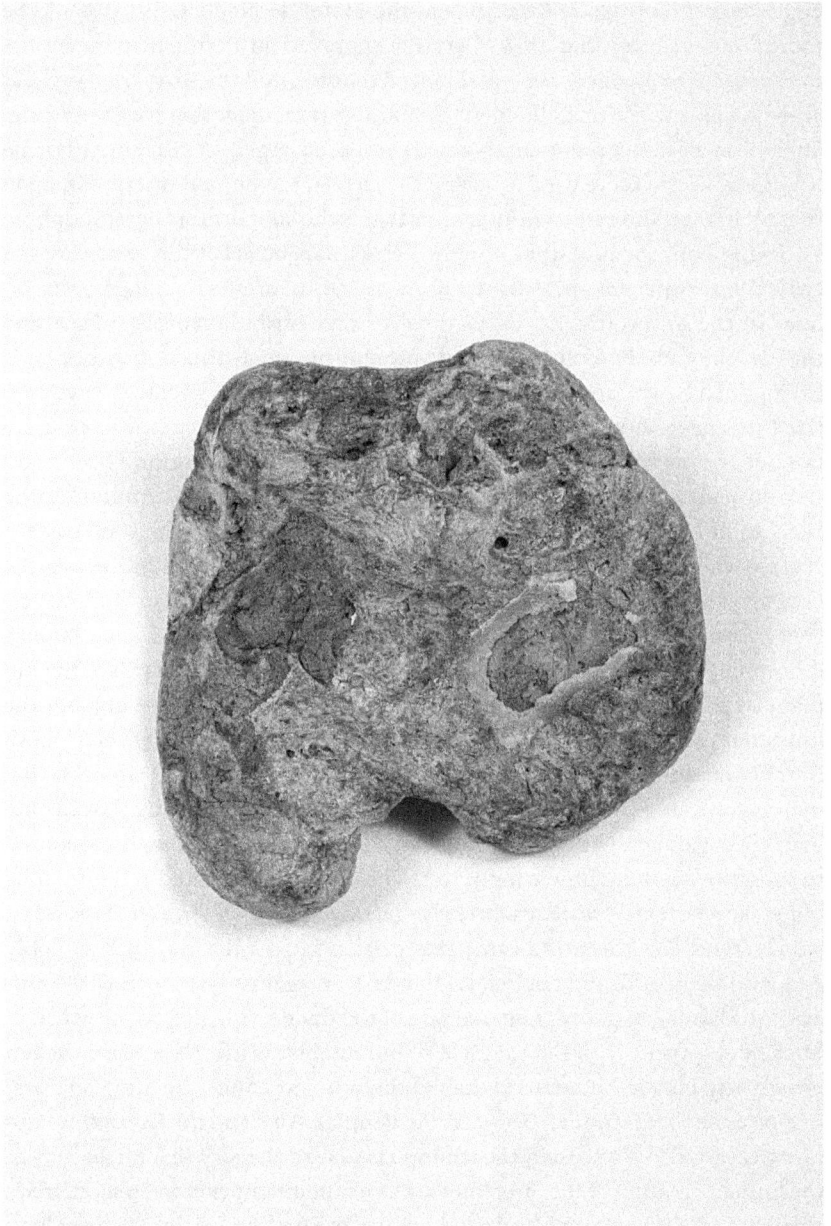

FIGURE 29. Hawaiian *Plastiglomerate*, 2013. This artwork/scientific study is the result of a collaboration between the artist Kelly Jazvac, the geologist Patricia Corcoran, and the oceanographer Charles Moore. Plastiglomerate is made from molten plastic debris and beach sediment, including sand, wood, rock, and shell. Photo credit: Jeff Elstone.

generally, I have been unable to find a single art-historical monograph about early modern art that problematizes in any extended way the connections between geology and the human designs wrought into stones—emphasis, generally speaking, remains on the cultural (the human, social, economic, etc.) aspects of art.[23] By choosing to see both what is carved into the limestone *and* the limestone itself, we look on with eyes from the Anthropocene. But, as Belleforest's map suggests, as do also material conditions for construction in early modern Caen and other elements we shall see in due course, humanists perhaps already looked on with similar eyes.

Many other limestone constructions sprang up in 1530s Caen, further sites for the interlocking of timescales via the human crafting of local exterranean matter. The most exemplary construction that signals this reconnection of geology and human-scale history is the Hôtel d'Escoville, commissioned by Nicolas Le Valois d'Escoville in 1533, which now houses the town's tourist office, turning this early modern lithic emblem, an earthwork for sure, into a contemporary showcase of Caen-ness.[24] Entering into the main courtyard, the early modern and modern visitor is, once again, most likely to *not* see the materiality because of the (human) architecture. Just as Jean-Luc Nancy argues that landscapes obfuscate the land they represent, so the human-wrought details on architectural structures can draw our attention away from the medium— as with the *trionfi*.[25] On the *factum* level, we see that the façade of the *aile des pavillons* features a central pavilion, atop which sits an intricately sculpted dormer window (*lucarne*), loggia on two levels, as well as a busy roof whose chimney and campaniles recall (in miniature form) the skyline-like roof of the Château de Chambord. On the façade of the *aile des statues* (Figure 30) one sees statues of David and Judith, bas reliefs depicting the Rape of Europa and the Freeing of Andromeda, as well as other features such as putti, garlands, and *rinceaux*. As with Petrarch's limestone triumphs, viewers (early modern and modern) obviously had and have the choice to see either the support, the inscription, or both, that is, either the figures (David, Judith, Europa, Andromeda) and the respective texts and traditions to which they belong or/and the limestone blocks employed here—especially their size. Indeed, demand for huge blocks of local stone without faults and fractures, the kind of large and more expensive pieces of stone required for the production of the David and Judith statues, meant that quarry sites shifted increasingly away from the center of Caen toward areas slightly farther away: Vaucelles, Fleury-Allemagne, and La Madrerie-Venoix.[26] In other words, from a material perspective, as the town witnessed more limestone being made exterranean and being inscribed by Italianate details, it was also experiencing a restructuring of its extraction practices.

FIGURE 30. *L'aile des statues*, Hôtel d'Escoville, Caen. © Phillip John Usher.

CHARLES DE BOURGUEVILLE

Trapped in our modern geological awareness, solicited in the Anthropocene to be cognizant of the exterranean, if we want to approach not just these early modern buildings but also early modern phenomenologies regarding the stoniness of Caen—and thus learn some lessons from the ways humanists might have been aware of the exterranean—we must turn from buildings to the writings of someone who lived among them, moreover in a time of crisis. Charles de Bourgueville (1504–1593), in his *Recherches et antiquitez de la province de Neustrie* (*Histories and Ancient Things of the Province of Neustria*) and subsequent *Recherches et antiquitez de la ville et université de Caen et lieux circonvoisins* (*Histories and Ancient Things of the Town and University of Caen and Neighboring Places*) (1588), shows himself to be not just a regional historian, trained in the methods of his time, but also a thinker of both lithic durability and—when put under human hammer—lithic impermanency.[27] As we shall see, Bourgueville's writings posit stone as a medium *and* encourage us to develop an ooltic and exterranean sensibility.

Much more than those of other Caennais writers of the sixteenth century, Bourgueville's writings place great emphasis on Caen's architecture, on its concrete materiality, and on its own very particular, almost cannibalistic, exterraneanity.[28] A first indication of this is that they contain numerous passages about the town's beauty and specifically about the stony nature of that beauty: Looking at Caen, one beholds, writes Bourgueville, "one of the most beautiful, most spacious, most pleasing and delectable [towns upon which one could hope to gaze]," a general appreciation to which he adds the following key point: The town is magnificent thanks to "its situation, the structure of its walls, its churches, towers, pyramids [steeples], buildings, tall houses [*pavillons*] and buildings, [and] big and wide roads," all made of locally extracted stone.[29] He instructs the would-be visitor that a good view of the town's buildings—in a slightly altered version of Homeric *teichoscopia*—is to be had from the city's bridges, whence one sees "the beautiful and sumptuous edifices, towers, pyramids, the castle, and houses, on both sides of the town."[30]

Going beyond such petrophilia, Bourgueville makes no secret of the fact that the beauty of Caen's buildings—as he perceives it—is directly related to the rock matter of which they are made. His text gets quite specific. At one point he notably describes the Saint Julian neighborhood as a place where can be found "quarries of the whitest, most polished, most gentle stone that one can find [and which] hardens when employed in construction." Superlatives thus impose themselves both for the town's beauty *and* the exterranean matter of which it is made. Bourgueville goes on to state that the particular quality of Caen's stone is that neither "time's abuse [nor] the force of the air or frost can harm it or damage it," adding that "the utility [*commodité*] of these quarries and the beautiful stones are one of the reasons the town is so finely adorned and well supplied with beautiful and excellent buildings, towers, pyramids, and other edifices."[31] Although today the quarries of the Saint Julian neighborhood are visible only in street signs (for example, one that reads "FOSSES ST JU-LIEN"), residents of early modern Caen would have, like Bourgueville, been able to appreciate more immediately the material genealogies of stone—the Saint Julian neighborhood was but a short walk from the historic town center.

Bourgueville's description of Caen's castle also accentuates—indeed, somewhat in the manner of the map in Figure 25—how castle/quarry/rock are intimately one. The castle, he writes, is "situated high-up, like a crown and a propugnacle [fortress] for this great town." It has moreover been "since time immemorial [*de tous temps*] one of the best of [the French] kingdom in beauty, size, and defensive might" for the specific reason that it is "seated [*assis*] on a natural rock."[32] Foregrounding the depth of the castle's and the dungeon's moats, where much of the rock for the castle would have been extracted, Bourgueville writes that they are "of astounding depth," to the point that "they could never be climbed [*ne sont subiets à l'escalade*]."[33]

Of late, our sympathy for hard and slow matter, as well as for the geological more generally, has to do with our realizing the consequences of extracting and burning fossil fuels. In sixteenth-century Caen, a similar "realizing" was occasioned by the sudden demand for huge and faultless blocks of limestone required to build new Italianate townhouses but even more so by the events of 1562, a calamitous year that marks the beginning of the French Wars of Religion that would carry on until 1598. The history of this period of civil unrest has been told in many ways, through lenses such as religion, politics (macro- and, increasingly, micro-), or gender, but never, as far as I know, via its relationship with matter, with geology. But what if—and indeed Bourgueville seems to do just this—human history (in which Protestants destroy limestone buildings) is *thought through* the stony and the exterranean? Bourgueville describes the worst of it as follows: "On Friday night and all day Saturday [May 8 and 9, 1562], all the temples, churches, and monasteries of this town were pillaged and sacked: Windows and organs were broken, [sacred] images were massacred, and all the churches' decorations that could be found were pillaged. [Everything] that was burnable was consumed by fire." Damages were estimated at "more than one thousand *écus*."[34] Many houses and buildings were attacked or demolished, as were various statues around the town.

The *Recherches et antiquitez* not only *record* the events of 1562 (in much more detail than I quote here); they also enact a variety of strategies for recovering or reproducing what was stolen or destroyed, that is, for problematizing the too-quick association of exterranean stone with durability. The book offers itself up as a site for the recovery of lithic loss, in several ways. The text includes the exact size of certain demolished objects, so as to keep a paper trace of the three-dimensionality of what once was made of local stone, for example, the "beautiful cross" mentioned on the map in Figure 25, where limestone was bought and sold.[35] Elsewhere, it registers the author's attempt to save destroyed stone monuments by creating new stone models. When a Protestant leader, the Duke of Bouillon, having taken over Caen's castle, feared that he might be attacked from the Collégiale du Sépulcre, he ordered the latter be destroyed, "having no respect," writes Bourgueville, for the fact that he would be reducing to rubble such "a beautiful temple and sacred place."[36] Not content with merely registering the event, Bourgueville also reacts as follows: "As they were getting ready to demolish this beautiful and ancient temple, I had a sketch [*pourtraict*] made of it." Bourgueville's plan, however, was not merely to produce a paper sketch but in due course to "have [that sketch] sculpted in stone," to be placed in the cemetery "so that our children, descendants, and successors might be able to see it and experience the loss of [*puissent . . . regretter*] the structure as I present it to them."[37] Bourgueville's phrasing catches our attention: The plan is specifically not just to have future generations remember the event of the *collégiale*'s destruction, not even just to enable them to see or to imagine it, but very specifically to have them experience the loss, to *experience the loss of*, the very structure that is represented before them, via a (miniature) stone reproduction. It is as if Bourgueville writes: *Look at this that you do not see! See the fragility of geomedia represented in this exterranean reproduction!*

Another tactic for representing-reproducing those stony parts of Caen destroyed by Protestants becomes apparent in the context of the destruction of the tombs of William the Conqueror and of his wife Matilda of Flanders: Their "two singular and magnificent sepulchers [as well as] their effigies, and lifelike representations [*au vif*], made in 3D [*taillees en bosse*], according to their true appearance [*selon le naturel*], were knocked down and demolished by these enraged hooligans" who had respect neither for the "royal and ducal" dignity of the deceased nor for the "five hundred years of antiquity" of the objects themselves.[38] To the destruction of these tombs that took place within the Abbaye-aux-hommes and Abbaye-aux-dames that William the Conqueror and Queen Matilda founded, the *Recherches et antiquitez* offer a direct material response by reproducing the royal couple's epitaphs (in Latin), to which Bourgueville adds his own French translations thereof (Figure 31). An example of the period's vogue for transforming stony epigraphy into the papery and the portable, Bourgueville's quotation of the royal epitaphs is also a rejoinder to the frangibleness of stone.[39]

This overriding concern for stone's endurance, aeonic or otherwise, is enunciated as early as the book's materially anxious preface: "Even if great kings and princes have performed memorable and generous acts, they still [need], or even necessitate, that learned and virtuous men of their own time must write down [*reduit par escript*] their heroic deeds" for fear that "memory of [these deeds] be lost with time, which devours everything."[40] To make his point, he invokes Alexander the Great, who "on seeing the ruins of Troy [said] that Achilles was very lucky to have Homer as the trumpeter or advocate [*preconiseur*] of his deeds." To drive his point home further, he enumerates other classical heroes who have benefitted from the durability of the written word: Aeneas, Achilles, Hector, Paris, Agamemnon, etc. Moreover, having produced, in 1556, a French translation of another account of the Trojan War—Pseudo-Dares Phrygius's *Excidio Troiae* (1556)—Bourgueville clearly makes a statement here that must be understood as referring to himself as just such a preserver of memory in the face of fallible material memory and that he here extends to the town of Caen itself. I saved mention of this preface until this late point in the chapter because Bourgueville's words could all too easily be taken as

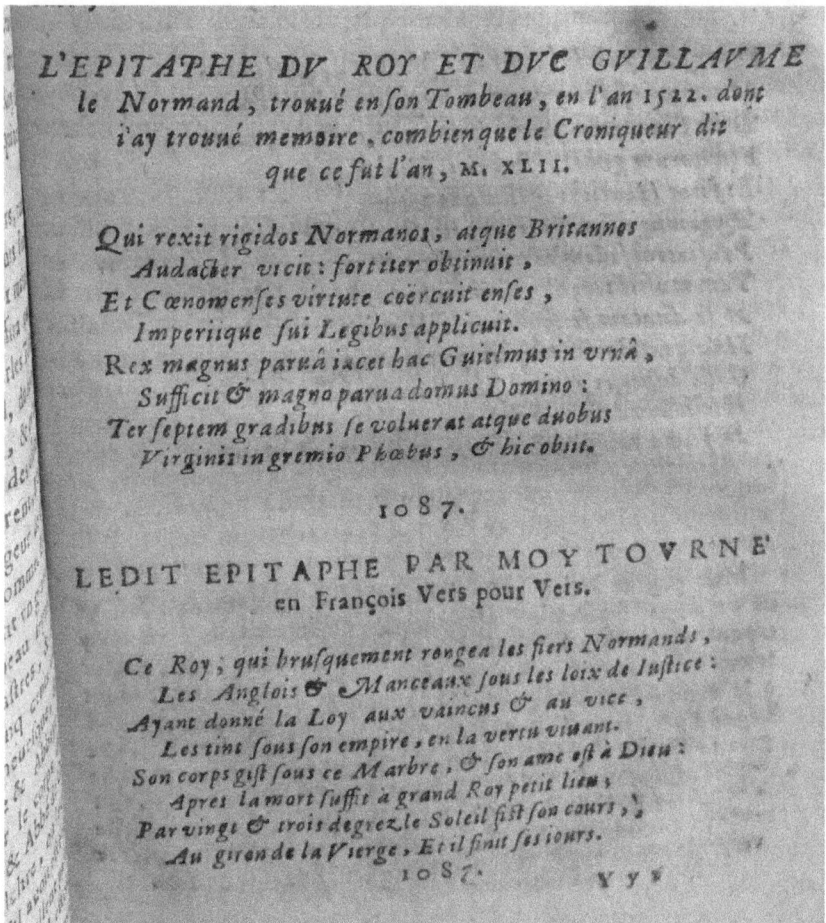

L'EPITAPHE DV ROY ET DVC GVILLAVME le Normand, trouué en son Tombeau, en l'an 1522. dont i'ay trouué memoire, combien que le Croniqueur dit que ce fut l'an, M. XLII.

Qui rexit rigidos Normanos, atque Britannos
Audacter vicit: fortiter obtinuit,
Et Cœnomenses virtute coërcuit enses,
Imperiique sui Legibus applicuit.
Rex magnus paruâ iacet hac Guielmus in vrnâ,
Sufficit & magno parua domus Domino :
Ter septem gradibus se voluerat atque duobus
Virginis in gremio Phœbus, & hic obiit.

1087.

LEDIT EPITAPHE PAR MOY TOVRNE' en François Vers pour Vers.

Ce Roy, qui brusquement rengea les fiers Normands,
Les Anglois & Manceaux sous les loix de Iustice:
Ayant donné la Loy aux vaincus & au vice,
Les tint sous son empire, en la vertu viuant.
Son corps gist sous ce Marbre, & son ame est à Dieu :
Apres la mort suffit à grand Roy petit lieu,
Par vingt & trois degrez le Soleil fist son cours,
Au giron de la Vierge, Et il finit ses iours.
1087.

Y y

FIGURE 31. William the Conqueror's Epitaph in Charles de Bourgueville, *Recherches et antiquitez*, II, 173. The Fales Library & Special Collections, New York University.

self-promotion, as a defense of the historian's trade, rather than as a genuine anxiety about our human investment in lithic stability, in the ability of the exterranean to tell us of our past and to conserve our present.

Lest the reader underestimate the importance of this part of his project, Bourgueville moreover makes this concern for exterranean durability consubstantial with his name: "I began this undertaking specifically to serve posterity, and to serve as a stimulus [*eguillon*] for the studious youth in order that they might have this kind of vigilance for the future, so that we might predict and prevent [*preuoir et preuenir*] future events and dangers which can indeed occur, for GOOD FORTUNE ERASES FORGETTING."[41] In other words: The *Recherches et antiquitez*, rather than presenting a survey of Caen's past, as one might expect from a historian, seek to preserve (whatever it can of) its present for the future—and what impedes or exhausts forgetting is *l'heur de grâce* (which we might render as "good fortune"). Now, the end of the previous sentence, which reads in French: *L'HEVR DE GRACE VSE L'OUBLI*, is an anagram of CHARLES DE BOURGUEVILLE—as a marginal side note points out lest the inattentive reader miss it. As such, the preface thus fashions Bourgueville as this "good fortune" that "erases forgetting," as an author who, comparable to Homer or the Dares he translated, will provide a material form that will preserve Caen better than the town's own stone ever could. Bourgueville claims for himself and his book this lithic agency.

Such author anagrams were of course common practice at the time, but the present one is striking for the way it encapsulates a literary project, that of writing as the only matter that can preserve memory.[42] The anagram is also printed around Bourgueville's portrait, below which appear verses signed I. V. D. L. F. (Bourgueville's compatriot and son-in-law Jean Vauquelin de la Fresnaye), which restate the "vibrant" nature of images and words: "This portrait and many a book / In painting and in writing, / Will cause to be seen again [*feront revoir*] and to live again [*vivre*] / Your face and your mind [*esprit*]."[43] While the city's architecturescape built from stone drawn from local quarries can suffer damage at the hands of Reformers, the printed page—*Ceci tuera cela!*—is capable of producing new life, of sustaining that whose solid existence is, in reality, limited in time. The work celebrates Caen as a petrous town and writes human history into the region's geology, but it also acknowledges just how fragile stone structures and lithic memory can be and offers up a paper memory in its place.

I have attempted in this chapter to read early modern Caen and its human history with an emphasis on the stony and the ooltic. Early modern inhabitants of the town did not just live *in the shadow* of William the Conqueror's huge stone castle, like actors on a film set. It was their history and their present, a reassuring continuity fashioned from stone extracted in or near their town. As the map in Figure 25 and Bourgueville's texts make clear, the relationship between stone quarry and stone building was visible, tangible, reapplicable, perhaps making the Italianate details on the Hôtel de Nollent and the Hôtel d'Escoville *not* efface the material of which they are made. If it is now the tourist office, is it as a site of reception for Italianate architectural trends? Or because it is made of local stone? Or both? What is clear is that early modern Caennais lived in dialogue with timelines and lithic entities much bigger than the human, even if the discipline of geology did not yet exist to undergird that move scientifically. They saw new buildings be built in local stone—and old ones torn down in the context of religious and civil upheaval. The townhouse builders, the angry Protestants, and a humanist-trained local historian already intuited in different ways that, when human history interacts with stone whose origins are literally rooted in the region's own grounds, that history requires—or rather just *is* a form of—geomedia, resulting in an intimate exterranean staining of the human.

CHAPTER 7

SALINE INTIMACIES

This chapter turns from limestone to salt. Like limestone, made mostly from calcium carbonate ($CaCO_3$), sodium chloride ($NaCl$), whether extracted from underground ("natural salt") or recuperated by evaporation of salt water ("artificial salt"), is—as modern science tells us and as early moderns perhaps, as we shall see, intuited—always ultimately exterranean: Its Na has its origins in sedimentary rocks that once formed the beds of ancient oceans, or else in rocks over which rivers ran on their way to the sea; its Cl once spewed from volcanoes and thermal vents.[1] Unlike limestone, the exterranean intimacies that it makes available are primarily chemical, that is, having to do not with what salt *is* or with human longing for a share in certain perceived characteristics (for example, aeonic endurance),[2] but with what it *does*. The fact that both limestone and salt are taken *ex terra* does not mean that humans live with them in the same way. Exterraneanity in and of itself does not determine how we perceive, access, use, or otherwise interact with a given matter. This chapter asks similar questions to the previous one, especially regarding what becomes of exterranean matter once it is materially separated from the Earth and the extent to which its previous connection can become perceptible, but it must end up being in many ways quite different. In the pages that follow, I shall thus explore a variety of early modern texts, objects, and images that, in different ways, address salt's exterraneanity and/or phenomenalize its intimacies, both benign and hostile, with humans.

VERBING 1: LISTS

Salt is primarily present, as already proposed, via what it *does*—and what it does it does because of its inherent chemical properties and how these cause it to interact with other things, living or dead, for example, the way that it draws water from other cells via osmosis. Since doing is the domain of verbs, it is not surprising that many texts, from the early modern period and today, strive to circumscribe sodium chloride in long (but necessarily not exhaustive) lists of what salt does. A recent mainstream TV documentary, *How Stuff Works: Salt*, begins as follows: "It can save you. And it can kill you. It can also conduct electricity, whiten your whites, and ward off the devil. It tastes like no other rock—and just try cooking a decent meal without it."[3] The journalist cuts the list short, concluding: "[Salt] has fourteen thousand known uses." Although the show's producers likely had no idea about this, this *verbing* was already how certain humanists wrote about salt. Take this list from the Huguenot author, potter, engineer, and ceramist Bernard Palissy, more specifically from his *Discours admirables de la nature des eaux et fontaines, tant naturelles qu'artificielles, des métaux, des sels et salines, des pierres, des terres, du feu et des émaux* (*Admirable Discourses on the Nature of Waters and Fountains, Both Natural and Artificial, as well as Metals, Salts and Salines, Stones, Earths, Fire, and Enamels*) (1580), a dialogue between "Practice" and "Theory"— Palissy's list is not that different from the one that opens the salt episode of *How Stuff Works*, even highlighting many of the same effects:

> Le sel *blanchist* toutes choses; le sel *endurcit* toutes choses; il *conserve* toutes choses; il *donne saveur* à toutes choses; c'est un mastic qui *lie et mastique* toutes choses; il *rassemble et lie* les matieres minerales; et de plusieurs milliers de pieces il en *fait une masse*. Le sel *donne son* à toutes choses; sans le sel nul metal ne rendroit sa voix. Le sel *resjouit* les humains; il *blanchist* la chair, donnant beauté aux créatures raisonnables; il *entretient l'amitié* entre le male et la femmelle, à cause de *la vigueur qu'il donne* és parties genitalles; il *aide à la generation*; il *donne voix* aux creatures comme aux metaux. Le sel fait que plusieurs cailloux pulverisez subtilement, *se rendent en une masse* pour *former* verres et toutes especes de vaisseaux; par le sel on *peut rendre* toutes choses *en corps diafane*. Le sel *fait vegeter et croitre* toutes semences.[4]

> (Salt *whitens* all things; salt *hardens* all things; it *conserves* all things; it *gives flavor* to all things; it is a gum that *connects together* and *glues* all things; it *sets and connects together* all mineral matters; and of many thousands of pieces, it *makes one mass*. Salt *gives sound* to all things; without salt, no metal would *have a voice*. Salt *delights* humans; it *whitens* the flesh, *giving beauty* to reasonable creatures; it *fosters love* between the male and the female, because of the *vigor that it gives* to genitals; it *helps* in procreation; it *grants a voice* to creatures as to metals. Salt means that several pebbles, smashed into small pieces, *come together as one mass* in order *to form* glass and all sorts of vessels; with salt, one can *give* to all things a *diaphanous* body. Salt causes all seeds *to bud* and *to grow*.)

The voice in Palissy's text is that of Practice, not that of Theory.[5] Practice does not say what salt *is* but rather enumerates these different actions that it performs. In most cases, salt is the active subject of the sentence. It is noteworthy, too, that there is a flattening of object types—the fact that salt *hardens* or *whitens* all things is not signaled as being valued differently from the fact that it also *delights* humans or *gives beauty* to reasonable creatures. Salt is, in a sense, everywhere and ready to interact with all things including humans, but it is not presented here as being primarily *for* humans. Whatever (human or nonhuman) gets in salt's way will be affected by it. Moreover, beyond what the individual items signify and beyond the flattening, the list format itself says a lot about salt—the list cannot enumerate *all* the things that salt *does*; it can only ever list some of them, as a specimen that allows the reader, as Umberto Eco has put it, "to imagine the rest." The list "transcends its own finitude" to say not just that salt *does* A, B, or C but that it does *lots* of things that cannot be said here, defining salt as an exterranean matter that calls for ever-renewed verbing.[6] The fact that any list of what salt *does* is only ever, ultimately, a metonymy for the enumeration of all salt's actions determined by its chemical composition is corroborated by the way Palissy and his contemporaries create other versions of the list we just quoted. Palissy, for one, repeats his list, with slight variations, at another point in his *Admirable Discourses*:

Le Sel *blanchist* toutes choses.
Et *donne son* à toutes choses.
Et si *fortifie* toutes choses.
Et si *est compagnon* de toutes natures.
Et si *entretient l'amitié* entre le masle et la femelle.
Et si *aide à la generation* de toutes choses animees et vegetatives.
Il *empesche la putrefaction* et *endurcist* toutes choses.
Il *aide à la veüe* et aux lunettes.
Sans le sel, il seroit impossible de *faire* aucune espece de *verre*.
Toutes choses se peuvent *vitrifier* par sa vertu.
Il *donne goust* à toutes choses.
Il *aide à la voix* de toutes choses animées, voire à toutes especes de metaux, et instruments de musique.[7]

(Salt *whitens* all things.
And *gives sound* to all things.
And it *fortifies* all things.
And it *is the companion* of all things.
And it *fosters love* between the male and the female.
And it *helps* all things, animate and vegetable, to *procreate*.
It *stops putrefaction* and *hardens* all things.

It *helps eyesight* and with glasses.

Without salt, it would be impossible to *make* any kind of *glass*.

All things can be *vitrified* thanks to its properties.

It *gives taste* to all things.

It *helps grant a voice* to all animate things, even to all kinds of metals, and
to musical instruments.)

Palissy's two lists have a number of items in common (that salt *whitens*, that it
fosters love between men and women), but there are also some items that appear
in only one of them (only the first list mentions how salt *hardens* all things).
This tension is further sharpened by one of Palissy's contemporary admirers,
the poet André Mage de Fiefmelin, the author of *Le saulnier* (*The Salt Worker*).
At one point in his poem—on which more in due course—Fiefmelin translates
Palissy's lists into alexandrines (twelve-syllable lines), the metric form recently
restored by the likes of Jean Lemaire de Belges, Clément Marot, Jacques Pele-
tier du Mans, and, especially, Pierre de Ronsard.[8] The French reads as follows:

> C'est luy qui blanchit tout, l'endurcit, le conserve,
> Luy donne estre et saveur, de sa fin le preserve.
> C'est le mastic qui lie, et mastique, et rejoinct
> Les pieces en leur masse et les pareils conjoinct.
> Ce sel donne à tout son, l'airain sans luy ne sonne :
> Aux creatures voix comme aux metaux il donne,
> Resjouit les humains, et blanchissant leur chair
> La beauté leur accroist, rend l'un à l'autre cher.
> Maintient en amitié la femelle et le masle
> Par la vigueur qu'il donne à la part genitale,
> Qui l'un et l'autre excite à generation.
> Toute semence croist par sa simple action,
> Tout se peut par luy rendre en un corps diaphane.[9]

To appreciate the additional work done by the very form of the alexandrine,
the way in which that form further materializes salt's verbing, any translation
must make at least some attempt to show the formal structure of the lines. One
possible rendering, which splits each line at its caesura and where *I*s remind of
the syllable count in the French original, replacing a sonorous structure with
a visual one, is as follows:

I I I I I I	I I I / I I I
It is salt that whitens everything	hardens it, preserves it,
I I I I I I	I I I I I I
It gives everything being and flavor,	protects it from its death.

I I I I I I	I I I / I I I
It is the glue that binds,	and which glues, and connects.
I I I I I I	I I I I I I
This salt grants sound to everything,	without salt bronze is silent.
I I I I I I	I I I I I I
To creatures it gives voice	as, too, to metals.
I I I I I I	I I I I I I
It delights humans	and, whitening their flesh,
I I I I I I	I I I I I I
Increases their beauty,	making them dear to one another.
I I I I I I	I I I I I I
Salt keeps bound in amorous friendship	the female and the male
I I I I I I	I I I I I I
Via the vigor it gives	to the genitals
I I I I I I	I I I I I I
By which one and the other are excited	toward procreation.
I I I I I I	I I I I I I
All seeds grow	via its action alone
I I I I I I	I I I I I I
All can, thanks to it, be made	into a diaphanous body.

If Palissy's lists thrive on and communicate a tension between the finite and the infinite, and if they thus invite the reader simultaneously to apprehend and to see beyond the individual *doings* they enumerate, to see also the more general principle that salt verbs, then Fiefmelin's alexandrine version of those lists takes this even further. The alexandrine is a syllabic data structure, here used to enumerate briny *doings*, and its horizontal fixedness in opposition to the ongoing addition of new lines brings home to the reader how salt's abilities can be apprehended *and* how that list could go on forever—the horizontal format is fixed but not the vertical one. To talk about salt's taste or to define salt by the ways in which we perceive it would be to play a Kantian game, for it would imply what Quentin Meillassoux—and others since—call correlationism; that is, salt only has meaning because *I*, as a royal Cartesian subject, taste it on my tongue, and it only exists *for me*.[10] The lists studied here, even if humans are present at times, go beyond this, for, both in how they describe salt's effect on metals and various nonhumans and in their very listy-ness, they propose as if scientific experiments a series of dissociated objects at the moment of their interactions (salt/metal, salt/flesh, salt/skin, salt/animals, salt/everything), interactions that do not necessarily depend on human perception.

VERBING 2: SALT MAN

We have just seen lists apprehending salt's (in)finite array of *doings*, which collectively capture both salt's fundamental chemical properties and some of the ways salt becomes present to other things, whether living or dead, human or nonhuman. A related *verbing* occurs when salt's doingness is ascribed to a literary character, as is the case of Pantagruel. Here, salt's agency is apprehended via a specific instrumentalization thereof. In this particular case, Pantagruel weaponizes salt by putting one of its characteristics to work in order to sow discord; that is, he hurls salt at people, with the primary goal of having them swallow it, become thirsty, and drink. He takes aim, in particular, at drunkards to make their situation worse still. His power comes from how he throws salt, just as Spider-Man's agency materializes in how he shoots spider webs from his wrists. This primary characteristic can all too easily end up buried under mountains of commentary, but there is no reason that Rabelais could not have—quite fittingly—titled his *Horrible and Terrifying Deeds and Words of the Very Renowned Pantagruel* (*Horribles et épouvantables faits et prouesses du très renommé Pantagruel*) (c. 1532) simply *Salt-Man*. To appreciate this fact, we must recall that Pantagruel was, before Rabelais came on the scene, a character in various fifteenth-century texts. In the *Life of Saint Louis* (*Vie de Saint Louis*), Pantagruel recounts his salty exploits as follows:

> Je vien de la grande cité
> De Paris, [et j'] y ay esté
> Toute nuit . . .
> A ces galanz qui avoyent beu . . .
> Tandis qu'ilz estoyent au repos,
> Je leur ay par soutille touche
> Bouté du sel dedans la bouche
> Doucement, sans les esveiller;
> Mais, par ma foy, au resveiller,
> Ilz ont eu plus soef la mitié
> Que devant.[11]

> (I come from the great city
> Of Paris—and I was there
> All night long . . .
> Pursuing these gentlemen who had imbibed . . .
> While they were resting
> I—with the subtlest approach—
> Threw into their mouths

(Quite carefully) salt, without waking them;
But—I swear to God!—when they woke up,
They were more thirsty indeed
Than before.)

Pantagruel's role here is that of Salt-Man, of he who *makes boozers thirsty* ("al-térer les buveurs") thanks to the thirst-inducing properties of salt.[12] He is the wielder of a chemical. In this text, the "drunkards of Paris" have to deal with the "bad tricks" of Pantagruel.[13] We find exactly the same salt-wielding Pantagruel in a more-than-sixty-thousand-line-long work by Simon Gréban, the *Mystery Play about the Acts of the Apostles* (*Mystère des actes des apôtres*) (c. 1460–1470). In that play, Pantagruel is a Breton devil whose job it is to throw salt into the mouth of sleeping drunkards.[14] In Gréban's *Acts* it is Lucifer who narrates Pantagruel's briny superheroism:

Panthagruel
Qui de nuyct vient gecter le sel,
En attendant autres besongnes,
Dedans la gorge des yvrongnes

(Pantagruel,
At night, stalks about throwing salt
—While he waits for other tasks—
Into the mouths of lushes)[15]

Rabelais's *Pantagruel* (1532) obviously turns this devil into a much more fully developed character—he is no longer *just* a salt-wielding devil, a secondary character in somebody else's story. He is now a giant, and—a first in the history of giants—he is, more or less, a likeable hero, not an incarnation of a cultural or political Other.[16] In the book's culminating chapters, however, in which Pantagruel and his men wage war against the bellicose enemies who have invaded his father's country, Pantagruel is again—indeed more than ever—Salt-Man. To begin, the invading enemies are Pantagruel's traditional adversaries: drunkards, or "the thirsty ones," here referred to as the "Dipsodes" (from the Greek δίψα, *thirst*). Their king is King Anarchus, King Anarchy. The battle that ensues is an augmented version of Pantagruel's defining gesture of throwing salt into the mouths of sleeping drunkards. Four key moments in the text develop Salt-Man's weaponizing of sodium chloride.

First, Pantagruel instructs a captured Dipsode to return to King Anarchus. Before telling him to head out, Pantagruel "gave him a box filled with euphorbia-resin and grains of *daphne gnidium*, steeped in brandy to become a syrup" ("luy bailla une boette pleine de Euphorge et de grains de Coccagnide confictz en eau ardente en forme de compouste"), further instructing him to tell King Anarchus that "if can swallow just one ounce of this mixture without needing a drink, he could stand up to Pantagruel without any fear" ("s'il en pouvoit manger une once sans boire . . . il pourroit à luy resister sans peur").[17] In the challenge that Pantagruel sends to King Anarchus with the released prisoner of war, the role of salt is taken by two other chemical compounds: euphorbium, a milky juice extracted from several species of the cactus-like Euphorbia, a substance known to the ancients, formerly used as an emetic and purgative, and still found in brands of nasal spray;[18] and a syrup made of the crushed berries of the highly poisonous *Daphne gnidium* evergreen shrub. Like a video-game character who has picked up an extra weapon, Salt-Man wields new chemicals to induce thirst: "As soon as [King Anarchus] swallowed one spoonful of the mixture, he suffered such a burning inflammation of the throat, with an ulceration of his uvula, that his tongue peeled off" ("tout soubdain qu'il en eut avallé une cueillerée, luy vint tel eschauffement de gorge avecque ulceration de la luette, que la langue luy pela"). As we would expect, this caused him to drink: "As a remedy, he could find no cure except that of drinking without pause, for as soon as he withdrew the goblet from his lips, his tongue burned" ("pour remede qu'on luy feist ne trouva allegement quelconques, sinon de boire sans remission: car incontinent qu'il ostoit le guobelet de la bouche, la langue luy brusloit").

Second, and returning to his primary saline weapon, Pantagruel gets ready for battle as follows: "He took their ship's mast in his hand as though it were a pilgrim's staff and stowed in its crow's nest the two hundred and thirty-seven casks of white Angevine wine left over from Rouen, and strapped to his belt the boat's hull crammed full with salt" ("[il] print le mast de leur navire en sa main comme un bourdon et mist dedans la hune deux cent trente et sept poinsons de vin blanc d'Anjou, du reste de Rouen, et atacha à sa ceincture la barque toute pleine de sel"). Readying himself to war thus means picking up a salt-filled ship and much wine. Third, Pantagruel uses that reserve of salt to begin war: "Pantagruel began to sow the salt that he had in his ship, and because the enemies were sleeping with their jaws gaping wide open, he so filled up their gullets that those poor wretches began barking like foxes, crying 'Ha! Pantagruel, you're heating up our firebrands!'" ("Pantagruel commenca semer le sel qu'il avoit en sa barque et, par ce qu'ilz dormoyent la gueulle baye et ouverte, il leur en remplit tout le gouzier, tant que ces pauvres haires toussissoient comme regnards, cryans: 'Ha! Pantagruel, tu nous chauffes le tizon!'"). Having poured salt into his enemies, Pantagruel urinates profusely, causing a veritable flood whose saltiness cannot be overlooked—certain of those present interpret the event as if "sea-gods such as Neptune, Proteus, Triton, and the others were persecuting them and that it was in fact sea-water and salty" ("que les dieux marins Neptune, Protheus, Tritons, aultres le persecutoient et que, de faict, c'estoit eaue marine et salée").

Fourth, when Pantagruel attacks the enemy giant Loup Garou (French for "Werewolf"), he encounters an avatar of the traditional open-mouthed drunkard: "Pantagruel, seeing Loup Garou approach with his chops agape, bravely advanced towards him and yelled as loud as he could, 'Death, you scoundrel, death!'" ("voyant Pantagruel que Loupgarou approcheoit la gueulle ouverte, vint contre luy hardiment et s'escrya tant qu'il peut: 'A mort, ribault, à mort!'"). As we would expect from this Salt-Man, Pantagruel's first battle blow involves stuffing his enemy full of salt: "From the boat which he bore on his belt he threw eighteen barrels and one Greek pound of salt at Loup Garou, stuffing his gullet, nose, and eyes with it" ("lui jetta de sa barque, qu'il portoit à sa ceincture, plus de dix et huyct cacques et un minot de sel, dont il luy emplit et gorge et gouzier, et le nez et les yeulx"). Much more could obviously be said about the role played by salt in the works of Rabelais.[19] Let us remember, here, how Bakhtin reminds us that Rabelais wrote *Pantagruel* as France was experiencing "a long and intense spell of hot weather and drought" (approximately March to September 1532), a situation that gave rise in France to a generalized perception of "cosmic terror."[20] This would surely make Pantagruel both hero and villain, he who brings thirst in a way 1530s readers would have related to—but he indeed wields salt's power only against his enemy. Whereas Palissy and Fiefmelin point to salt's chemical properties via verbing lists that transcend their own finitude, hinting further at how salt acts equally on humans and nonhumans, Pantagruel, both before and in the writings of Rabelais, shows salt's chemical agency as instrumentalized within the limited bounds of specific narratives. In his own way, Pantagruel becomes a geologic force, wielding weaponized salt to wreak havoc.

EXTRACTION 1: NATURAL VERSUS ARTIFICIAL SALT

In the above sections on what salt *does* no mention was made of where salt comes from, of its exterranean origins. Such origins, indeed, can be easily forgotten—just as it is all too easy to forget that our ride on the bus requires the extraction of petroleum or natural gas. The connection of extracted matter to Terra is never automatic—it is always an act of deliberate connecting. For such connections we must thus turn elsewhere. As noted at the beginning of this chapter, for early modern humanists salt was either natural (mined) or artificial (produced by evaporation of salt water). A first, obvious, point is thus that it is the first category that is, directly, exterranean and that, even before the eighteenth-century recognized that salt suspended in water had originally eroded from rocks—and was thus also exterranean—the hierarchy making extracted salt natural already emphasizes salt's extraction *ex terra*. Whether or not individual writers or artists fully perceived this, it is clear that early moderns thought of salt as emerging from both earth and water. In his *Traité du feu et du sel* (*Treatise on Fire and Salt*) (1618), Blaise de Vigenère writes—after Agricola[21]—that there is "natural" salt (*sel naturel*) that "grows like icicles, or *as rock* inside the Earth" ("croise en glaçons, ou en roche à par soy dans la terre") and "artificial" salt (*sel artificiel*) that "is made from sea water, or from salt liquor, like a brine, which is drawn from salt wells, as in Lorraine, Free County of Burgundy, which is then boiled and frozen in fire" ("se fait de l'eau de la mer, ou de la liqueur, comme vne saumere qui se tire des puits salins, ainsi qu'en Lorraine, & la Franche-comté de Bourgogne, qu'on fait décuire & congeler sur le feu").[22] But it is not just in early modern treatises about salt that the origins of salt become perceptible.

Let's imagine a possible scenario for an awareness regarding salt's exterraneity: King François Ier of France (r. 1515–47) sits down to dinner, on a day when a foreign ambassador is visiting Fontainebleau, perhaps that of Holy Roman Emperor Charles V. He looks out at his guests and invokes a popular proverb of the time: "The feast that wanteth salt is fit for deuills" ("C'est vn banquet de diables ou il n'y a point de sel").[23] Next, to reassure his guests that they are not, in fact, devils, he orders a servant bring to the table the saltcellar or *saliera* that Benevenuto Cellini made for him in the early 1540s (Figure 32). Most likely placed between the king and the foreign ambassador, the *saliera*'s dish is filled with sodium chloride. As the king and the ambassador reach for salt, Ceres and Neptune look back and remind them: Salt can come from the sea or from the Earth. Moreover, we note that Cellini's Neptune might look confident, but Ceres holds her left breast, as if to indicate that *she* is the productive one here, and her associations, clearly, are with the Earth.[24] Not all early modern saltcellars provided such direct awareness of salt's exterraneity. A good number of them took instead the shape of ships (*nefs*), thus equating salt only with the sea. This too had been Plutarch's message in his *Symposiacs*: "Among all those things the Earth yields, we find no such things as salt, which we can have only from the sea."[25] But when Blaise de Vigenère quotes the *Symposiacs* in his *Treatise on Fire and Salt*, he leaves out Plutarch's statement on salt's oceanic origins, citing only what comes next: "Nothing would be edible without salt, which mixed with flour seasons bread" ("Sans le sel rien ne se peut manger d'agreable au goust, car le pain mesme en est plus sauoureux si on y en mesle").[26] In other words, he explicitly suppresses mention of how salt comes from the sea. Cellini's *saliera*, in any case, seems to anticipate precisely the later scientific knowledge that salt always originates *ex Terra*.

FIGURE 32. Benevenuto Cellini's *saliera* (c. 1540). Kunsthistorisches Museum, Vienna, Austria, Inv. 881. Erich Lessing / Art Resource, NY.

For confirmation of the potential blurring of the artificial/natural boundary, of the fact that it is quite justified to talk about salt's wet exterraneanity, we can turn briefly to a largely forgotten poem by a Calvinist who lived and wrote on the island of Oléron off the Atlantic coast of France, André Mage de Fiefmelin's *Le saulnier, ou de la façon des marois salans et du sel marin des isles de Sainctonge* (*The Salt Worker; or, The Ways of the Salt Evaporation Ponds and on Sea Salt from the Saintonge Islands*) (1601), from which we already quoted a verse version of Palissy's salt list.[27] The *Saulnier* is an unusual poem. It offers a sort of verse topography of the Saintonge islands and their salt evaporation ponds (Figure 33), complete with lots of technical vocabulary, squeezed into alexandrines, as well as a tale of the poet's own presence within that space: He walks about, happens upon an older and experienced salt worker with whom he has a salty lunch, etc. Although the poem is closely focused on the production of what Agricola and Blaise de Vigenère call "artificial" salt, and even though the emphasis throughout is on the tools and procedures for this *wet* production, the reader nonetheless witnesses a definite tension between salt as (simply) exterranean or (simply) watery, in terms of both *what* the poem says and *how*.

FIGURE 33. The evaporation ponds on the island of Oléron in the twentieth century.
© Phillip John Usher.

The poet starts in almost epic mode—and epic modes are rarely *pure*—by defining his subject negatively, saying that he will *not* sing "of that first invention / Which gave us crops under grain-producing Ceres [*Ceres la blediere*]," nor of Silenus (Bacchus's companion), nor of "What Bacchus does with his vinous grape [*en son raisin vineux*]."[28] Recognizing that salt can come from the Earth or from water (seas, fountains, lakes), and after a typology of different rock salts, the poet states that he will sing not of rock salt ("Non le fossile") but of salt produced in a specific watery place, that is, "the ponds [*marais*] encircled by the Saintongeois ocean [*qu'enceint l'Ocean Sainctongeois*]."[29]

Be this as it may, the poem frequently makes salt a function of the exterranean. The most obvious instance of this is when the poet makes an explicit comparison between the production of salt on the island of Oléron and the extraction of gold in the New World. The poet argues that the king reaps more profits from this "island mine" (*fonds insulaire*) than from the mines in the Americas because of the taxes assessed on salt sold through the royal storehouses (*greniers à sal*): "Ainsi croy-je que l'or qu'on tire de ses mines, / Qu'à grands frais et hazards et par flottes insignes / Du Peru on apporte au thresor de nos Rois, / Ne vault tant de rapport que ce Sel Saintongeois" ("The gold extracted from his mines, / Which at great expense, much danger, and which, by ship / Is brought from Peru to our Kings' treasury, / Is not as profitable as this salt from Saintonge").[30] Salt, then, is like gold, and the *fonds insulaire* and the *mines* of Peru are comparable sites of extraction, even if the first is watery. Turning immediately to matter itself, the poet compares "this singular fruit [*ce fruict singulier*]" that is French salt to "philosophical gold [*cest or philosophique*]," that is, the gold of alchemists like Giovanni Bragadino, again concluding that gold yields less profit than salt.[31] The poet makes these points to denounce the salt tax (*gabelle*)—but important here, if we focus on the relationship between tenor (*salt/evaporation ponds*) and vehicle (*gold/mines*) rather than on the political context, is that the two forms of extraction are considered analogous. Fiefmelin may here have been reacting to Catherine de Médicis's reported statement to her son, King Charles IX, during a visit to the salt evaporation ponds at Bayonne: "*Here are your gold mines . . .*"[32] At another point in the poem, the extraction of salt is likened to the mining of precious stones in India (crystals, pearls, and diamonds), such that salt, too, might be called a gemstone (*pierre gemmeuse*), a statement qualified by a subtle formula: This briny gemstone is "more than terrestrial, aqueous" (*plus que terrestre, acqueuse*). This is not an outright statement—salt is not terrestrial—but more of a hybrid: *More* than terrestrial, it is *watery*.[33] The formula is revived further on in the poem, when the poet says of salt that it is a "grain more watery [*plus aqueux*] than terrestrial [*terrestre*]."[34]

Another way in which the poem links salt extraction to the Earth is via its emphasis on *terroir*:

. . . à faire bon sel n'est bon tout territoire,
Tesmoin le sable et bris de couleur jaune et noire.
Ains le marin terroir, dur et doux comme estain
Qui, ferme séelle mieux, au sel est souverain.
La terre propre à pots, et tenante et visqueuse
Est celle qu'en marois, comme meilleure, on creuse,
Limitrose à la mer, plus basse en son platin
Que la mer, la Chenal, et le ruisseau voiysin.
Et faut que dedans soy ceste terre marnine
Ait semence de sel, que l'art à sel provigne:
Comme en l'or il y a quelque semence d'or.[35]

(. . . for making good salt, not all earth is suitable,
As shown by that sand and earth that is yellow and black.
But maritime *terroir*, hard and soft as tin
Which, firmer, more watertight, is perfect for salt.
Pottery-making earth, both sticky and viscous
Is that which, in the ponds, is dug,
Next to the sea, lower in its shoal
Than the sea, the channel, and the neighboring stream.

It is necessary that inside it this maritime earth
Contain some seed of salt, which salt art produces in abundance,
Just as in gold there is some seed of gold.)[36]

A specific kind of earth, then, or territory/*terroir* is necessary for the production of salt, a type of earth thought to contain some kind of "seed of salt," just as the hillside must contain some "seed of gold" if one is to mine there. We are in the domain of *terroir*, to which an early modern proverb speaks well: "Telle terre telle cruche," or "As the earth, so the pottery pitcher"; that is, the earth itself determines what it will produce.[37] This attention to the specific nature of the earth is close to the kind of earth-focused emphasis of practical texts about agriculture, especially Olivier de Serres's *Théâtre d'agriculture et mesnage des champs* (*Theater of Agriculture and Management of the Fields*) (1600). The *Théâtre*, which begins by affirming that "la Terre est la mere commune et nourrice du genre humain" ("the Earth is the common and nourishing mother of the human race"),[38] dedicates its first chapter to the topic of the knowledge of earth/ *terroir* (*De la cognoissance des Terres*):

> Le fondement d'Agriculture est la cognoissance du naturel des terroirs que nous voulons cultiver, soit que les possedions de nos ancestres, soit que les aions acquis: afin que par ceste adresse, puissions manier la Terre auec artifice requis; et employer à propos et argent et peine, recueillions le fruict du bon mesnage, que tant nous souhaitons: c'est à dire contentement auec moderé profit et h[on]neste plaisir. Par là doncques nous commencerons nostre Mesnage, et dir[on]s qu'on remarque plusieurs et diuerses sortes de terres, . . .

> (The foundation of Agriculture is knowledge of the natural properties of the *terroirs* that we want to farm, whether we inherited them from our forebears or purchased them, so that with this skill we might be able to handle the Earth with requisite skill, and thus properly use both money and labor and be able to collect the fruits of this good management as much as we want, which is to say happiness with a modest profit and honest joy. To this end, we thus begin this Work and will say that we notice that there are several different kinds of earth . . .)[39]

The salt workers of Oléron Island are thus following, in their attention to land/ earth/*terroir*, the kind of advice followed especially by those who cultivate the Earth/earth directly for its crops, casting watery *saliture* as exterranean, beholden to both Neptune and Ceres. At one point, the poem even pauses to comment on the fact that the muddy banks of the evaporation ponds are themselves fertile and cultivated by the salt workers: The "petits tertres fertiles / En legumes et bleds aux seuls Saulniers utiles" ("little fertile mounds / With their vegetables and cereals are used only by the Salt Workers").[40] The bread made from this "salt land" thus carries a particular and pleasing taste ("De ce terroir à sel prend sa saveur si souefve / Le pain qui, blanc, en vient").[41] It is not just to practical works of agriculture that Fiefmelin's *Saulnier* is close but also to the pastoral and the georgic, two modes that often overlapped in the early modern period.[42] Fiefmelin's *Saulnier* shares much with Ronsard's long decasyllabic poem *La salade* (c. 1568) and with Du Bellay's "Moretum de Virgile" (imitation of Virgil's *Moretum*), published in the *Divers jeux rustiques* (1559), for example in the details of the lunch that he shares with the elder salt worker.[43] Throughout, then, the *Saulnier* mixes registers, likening the work of the salt worker to that of both the miner and the farmer or pastoral wanderer. "Artificial" salt is as profit producing (for the king at least) as mined gold, but its cultivation requires the techniques and knowledge more of the farmer than of the miner, fashioning the process as a kind of wet agriculture, realigning the comparisons between mining and farming that we have examined elsewhere in this book.

EXTRACTION 2: LIKE MAN, LIKE SALT

A Neo-Latin poem published by the German humanist Conrad Celtis in 1502 provides a final glimpse at saline intimacies. Included in Celtis's *Quatuor libri amorum secundum quatuor latera Germanie* (*Four Books of Love, According to the Four Sides of Germany*), a curious book of poems linking Celtis's love for four women to four geographic regions, the poem describes how a human body descends into and ascends from the Polish salt mines of Wieliczka, just outside of Kraków.[44] Already significant by the thirteenth century or perhaps earlier, the mines of Wieliczka were, throughout the sixteenth century, *the* example of "natural or fossil and mineral [salt]" ("[sel] naturel ou fossile et mineral"), that is, specifically exterranean salt, as noted for example in Jean de Marconville's *De la dignité et utilité du sel* (*On the Dignity and Usefulness of Salt*) (1574).[45] The mines would later become the location for extravagant festivities such as the marriage of Augustus III of Poland to Maria Josepha of Austria in 1719, as shown in a fabulous *livre de fête*,[46] and they would subsequently receive extensive textual and graphic treatment in Diderot and d'Alembert's *Encyclopédie*. Specific to Celtis's poem, written during Celtis's stay in Poland in 1498–1490, as we shall see, is the way in which it likens human movement to that of extracted salt.[47]

Addressed to a named but unidentified friend, Janus Terinus, the poem begins by imagining that friend calling out to Celtis, asking him to descend into the mine: "Valeto, / Celtis, et infernas i, rediture, domos!" ("Farewell, Celtis, go down to the underworld . . . and then come back up!").[48] The descent into the salt mines is thus presented as a katabasis or, more specifically, as one commentator put it, as an *Abstieg in die Unterwelt*, descent into the underworld.[49] Such a qualification also implicitly suggests that the reverse movement—that is, the exterranean one—is an ascent from the underworld, that is, from a place of dead souls to the world of the living. This dead/living opposition is doubled by a second distinction between dark and light, which we should examine in more detail, for this is central to human/salt intimacy. The poem, in these ways, is thus strikingly different from the later engraving realized for the *Encyclopédie*, which, while it gives a clear sense of the interconnection of the underground mines and the world above, neither darkens the underground spaces nor mythologizes them in any way. Celtis first describes the experience of adjusting one's eyes upon descent into a dark place: On entering into the wide throat of the immense cave, he writes, the eyes cannot perceive any floor below—that is, until a torch (*fax*) is thrown down to provide light. Having provided light for the reader's perception, Celtis then describes what *could have* been seen previously, namely a horse-driven capstan-and-winch system whose ropes and pulleys allow miners—and *les curieux*—to descend. In the description of his descent, Celtis again foregrounds moments of perception: Being lowered down on a rope into the Polish salt mines is said to be like flying, and, as it descends, the body shakes ("ego sum tremulus toto cum corpore vinctus").[50]

Once Celtis has described the lowering of his body into the mines, he turns his focus to the darkness not only by asserting the absence of light but also by rendering as precisely as possible *what one sees when one sees nothing in the dark*: In the blind world of the salt mines, there are dark/gloomy/black stars, "stars that give birth to no light" ("quae nullum sunt paritura diem").[51] Looking into the dark expanses of the salt mines means seeing black stars, of viewing black on black, of seeing not-seeing. Celtis details the exact same experience as Milton would later describe in *Paradise Lost*, "No light but rather darkness visible."[52] Compared to the perception of low light, which relates to shadows and the shadowy, Celtis's description of the black stars of total darkness in the mines is allied rather with an acute hyperawareness of self.[53] Indeed, directly after this visual phenomenon, Celtis returns again to his trembling body. Paler than a cadaver—*we are thus offered a brief glimpse from outside*—Celtis describes himself, on the end of the rope, suspended between life and death ("inter vitam medius mortemque").[54] As he dangles at the end of the rope, deep inside the salt mine, Celtis says he fears knowing the same fate as Daedalus. As already hinted at by the opening address from his friend, Celtis's final exit from the mines is likened to exiting Tartarus, that dungeon of suffering from Greek mythology.[55] Thus, to summarize, Celtis is lowered into a mine—but the poem is about much more than that. As one scholar has put it, the reader's perception flickers between seeing the winch-and-pulley system that makes Celtis's journey possible as (a) modern technology (*moderne Technik*) and (b) as "eine unheimliche Höllenmaschine" ("an uncanny/strange infernal machine").[56] But both the classical references *and* the strange machinery that loses its pure status as technology in the way that Celtis depicts it as *what dangles him close to death* evoke first and foremost the phenomenological experience of *being in the salt mine*, representing the whole experience—which indeed seems traumatic—as a coming back to life.

There is, perhaps surprisingly, only one direct mention of salt in Celtis's poem—but it is an important one. Celtis writes that from these salt mines many rocks (*multa saxa*) are taken from the dark caverns ("de umbrosis cavernis"), which are in turn fired above ground to become a salt "whiter than snow [*niveum*]."[57] As François de Belleforest also explains in his *Cosmographie universelle*, when matter is first extracted it resembles black stones ("il semble que ce soient des pierres noires")—but once these black stones have been fired and crushed, the matter looks like white flour (*farine*) or white alums (*alum blanc*), that is, double sulfate salts.[58] In other words, if the poem offers a phenomenology of mining salt and not just the story of an individual's touristic descent into the mines, it is via this emphasis on how exterraneanization turns *black rock* into *white salt*, just as pale-Celtis-dangling-in-the-dark passes through the impossibly dark *Unterwelt* in order to reemerge into daylight. The comparison between sodium chloride and the human body is emphasized when Celtis writes of a specific fear, namely that were he to die, he would become—like the salt *before* extraction—one with *Terra*; he would even give his name to this mine ("mea . . . fatalem nomina terram").[59] He does not die, however—he reemerges, as fully exterranean as the white rock salt the mines were built to extract.

This chapter might also have discussed other early modern contexts, such as Austria's Salzbergwerk Dürrnberg, famous for its *sinkwerks*, or else Switzerland's Bex Salt Mines, or Italy's Volterra salt mines, or countless other sites, many of which Agricola includes in his cartography of salt in his *De natura fossilium*.[60] Salt mining was happening all over Europe and elsewhere in the early modern period. But the point here has not been to map out all early modern mining of salt or even to offer an exordium on such a topic. Rather, emphasis here has been on the (in)visibility of salt's exterraneity even when attention is being paid to its chemical agency. On the one hand, Palissy and Fiefmelin apprehend salt's verbing, that is, its (seemingly infinite) capacity for interacting with humans and nonhumans, whether or not humans are present to observe— but those lists themselves do not invoke where that salt comes from. For that one has to turn elsewhere, even if only to other parts of the same works. Rabelais and his predecessors personalize—and weaponize—salt's chemical agency, turning it into an instrument wielded by a devil-cum-giant, but here, too, the exterranean origins of salt, the fact that salt has to be extracted, is barely

FIGURE 34. Ecognostic jigsaw. © Phillip John Usher.

perceptible. On the other hand, Cellini's *saliera* (Neptune *and* Ceres), Fief-melin's poetic portrait of Oléron's salt evaporation ponds and concomitant blurring of the terrestrial/aqueous distinction, as well as Celtis's poem on his descent into and ascent from the mines at Wieliczka all foreground—or sug-gest—salt's exterraneanity. Salt, once extracted, is present primarily in what it does, but that doing's connection to the Terra from which it ultimately de-rives is only ever available sometimes, under certain conditions—we might think, indeed, of Timothy Morton's point that "ecognostic jigsaws are never complete" and further of the fact that the average saltshaker these days, un-like Cellini's cumbersome but awareness-provoking model, makes no claim to ecognosis.[61] The early-twentieth-century invention and addition to table salt of anticaking agents—magnesium carbonate ($MgCO_3$) or calcium silicate ($CaSiO_3$)—allowed saltshakers to become closed vessels, industrial, making it even less likely that its user would reflect on salt's origins.[62] For a renewed ex-perience of "ecological awareness" of the "dark-depressing" variety, one might reach instead for a real ecognostic jigsaw whose shapes, when put together, represent Cellini's *saliera* (Figure 34). Now a glimpse of Neptune's arm, now of Ceres's breast, now of . . . The slower, the more frustrating the jigsaw, the less one is given to spending one's time on jigsaws, the better the jigsawing will go.

EXPLICIT

The chapters of this book have offered up an extended unpacking of—and apology for—a neologism (the exterranean), a first attempt at rehabilitating humanism *for* the Anthropocene.

Such an enterprise as this one challenges Jedediah Purdy's opposition, in his recent and much discussed book *After Nature*, between the "humanist subjection of the living world" in which "only people have interests and value" and posthumanism's claim to be less "human-centered." To stick with humanism, Purdy argues, would mean erasing "a great swath of the experience, perception, and relationships—to places, other living things, and practices . . . that have formed environmental imagination," an "obliteration" he casts as "a humanist judgment about what is real—human interests, human projects—and what is not: relations with the nonhuman world, a sense that parts and places of that world have their own values and purposes." The chapters of *Exterranean*, as they open up the humanist library for its alertness to ecologies of extraction, has plotted how texts and images produced by early modern humanists actually provide—to paraphrase Purdy—swaths of experience, perceptions, and relationships that are peculiarly urgent for an awareness of the material and immaterial interconnections between the Earth and the "stuff" we take from it. To overstate the opposition between humanism and posthumanism in order to cast the latter as some kind of savior philosopheme is precisely to perform obliterations that deprive us of an athenaeum that swarms with critical perceptual affordances. To think that as we stumble into the Anthropocene we are somehow suddenly able to invent ex nihilo the theoretical and critical tools to think it smacks of the same hubris that got us into this position in the first place.

Returning to humanism has not been the goal, of course, but the means— and the need for finding new perceptual means feels even more compelling now as I begin this final version of my conclusion in July 2017 than it did when I started to write his book in 2015, when the *Guardian*'s "Keep It in the Ground" campaign was in full swing. Last week *New York Magazine* published David Wallace-Wells's article "The Uninhabitable Earth," which details various physical and sociopolitical consequences of global warming.[1] The article makes many compelling points—but, like the COP 21 document discussed in this book's Introduction, it silences connections between our planet and the "stuff" we take from it. On the one hand, Wallace-Wells uses the word *emission* a total of ten times, to refer on each occasion to the burning of fossil fuels that leads to output of CO_2 into the atmosphere. On the other hand, when he uses the word *extract* (only four times), it is *never* (not a single time!) to refer to the extraction that precedes emission. When the word *extract* is used it is to describe first the way in which plants breathe in CO_2, then subsequently to a potential solution to our predicament, namely the removal (extraction) of CO_2 from the atmosphere, that process of imitating plants called Carbon Dioxide Removal (CDR) or Carbon Capture and Storage (CCS). The role of exterranean extraction in the growing uninhabitability of Earth is never mentioned.[2] Wallace-Wells never makes the fundamental point that the Earth is becoming uninhabitable *because* of extraction from the Earth. It would appear to be much easier to imagine geoengineering the removal of CO_2 from the atmosphere than to imagine differently where that CO_2 originated or to see our increasingly uninhabitable Earth as one, quite simply, from which too much "stuff" has been chopped off, burned, and resituated above, around, and in us.

Another 2017 checkpoint for the need for the exterranean readings developed here can be found on the front page of the June 2 edition of the German tabloid *Berliner Kurier*, which featured a picture of Earth and a headline that reads: "Erde an Trump: Fuck you!" ("Earth to Trump: Fuck you!"). The cover ventriloquizes Earth in response to the decision taken by the American president, who deleted the section on global warming from the White House website as soon as he took office, to withdraw the United States from the Paris climate agreement: "We are getting out."[3] In response to Trump's extractivist "coal-rolling" hypermasculine swagger,[4] already apparent in his preelection "America First Energy Plan," which outlined a "vision" about extraction ("unleash America's $50 trillion in untapped shale, oil, and natural gas reserves, plus hundreds of years in clean coal reserves"),[5] a plan that commentators have called "dark" and "delusional" specifically for its rootedness in "a nostalgic compulsion aimed at restoring a long-vanished America in which coal plants, steel mills, and gas-guzzling automobiles were the designated indicators of progress,"[6] the German tabloid thus decided to give Earth a voice. But unlike Niavis's Terra in Chapter 1, for example, this *Erde*, as has been so common in environmental discourse since at least the 1960s, is seen from afar, a mere object glanced at from space, as if seen *from nowhere*—again, there is disconnection between Earth, extraction, and extracted matter. The newspaper's message is a strong one, an essential one—but, theoretically, it leaves us grasping for where and what this *Erde* is, for what becomes of it as we anthroturbate and bring exterranean matter into dirty clouds that we here glimpse as if from a spaceship.

As if responding to Wallace-Wells's article "The Uninhabitable Earth" and to the *Erde* on the cover of the *Berliner Kurier*, the three sections of *Exterranean* have attempted to grasp Earth via a multiplication of humanist but antianthropocentric perspectives that situate the planet not as a big sphere seeable from nowhere but as a whole that is present *right here*, under, around, and in humans and thus not detachable from humans. This book's first section—Terra Global Circus—examined three early modern Terrae not to adduce some (nonexistent) linear history of Earth but to examine some of the ways in which Terra and the exterranean produce (conceptually) one other. The effect of this first section is hopefully akin to the one produced by the twenty-first century play from which it borrowed its name.[7] Chapter 1's reading of Paulus Niavis's late-fifteenth-century *Judicium Jovis* demonstrated how Niavis's Terra draws on but also differs from multiple classical and medieval understandings of that same figure, ultimately demonstrating the difficulty—but also necessity—of imagining Terra as *neither* just a bodily form/planet *nor* only as a vitality infusing all of *physis* but as both—as both connected by Fortuna's consistently inconsistent wheel. Chapter 2 turned to a mid-sixteenth-century poem by Pierre de Ronsard, in which we saw a tension between a poetic (proextraction) fantasy of a Terre anxiously willing to offer up its gold-filled mines, the material genealogy of that same gold as detectable in mention of the auriferous river Aurence, and, finally, the poem's brief allusion to gold coming from "far off" (the New World). It is precisely this tension that structures the exterranean. Chapter 3, via readings of Sebastian Münster, Michel de Montaigne, Bartolomé de las Casas, and others, turned from the mythological Terrae of Niavis and Ronsard, whose geographies remained somewhat aloof, to the post-1492 globe, in order to explore the insistence on exterranean extraction in discussions of the suddenly global Terra and to ask how such a call might allow for recognition/incorporation of anthropological difference even as we see both globe and surface depths.

The second section of this book—Welcome to Mineland!—shifted emphasis from Terra to terra. In the spirit of the festival organized by Philippe Quesne at the Nanterre-Amandiers theater after which it is named, it offered two different perspectives on going underground. Chapter 4 took up Georgius Agricola's

De re metallica. Normally read as a technical manual, this early modern "Miner's bible" was explored here in a different light, as a resource for perceiving exterranean activity close up, on and under hillsides. By getting as intimate as possible with those places and spaces where extraction occurs, the aim here was both to unframe and to reanimate landscapes *and*, by seeing their liveliness, to see them in their connection to the Terra of Part I. In Chapter 5, to explore a different aspect of the antagonistic exterranean relationship between humans and the nonhuman *terra*, we descended once and for all underground into the galleries where miners extract material *ex terra*. Via three authors—Agricola, François Garrault, and Paracelsus—the chapter discussed the early modern belief in mining spirits (*daemones*). Rather than to provide a cultural history, and beyond arguing for the vibrancy of underground extraction points, Chapter 5 thus wanted to give voice to the *être-patte-de-taupe* or *mole-paw-being* that emerges from Quesne's play.

Section 3—Hiding in Exterranean Matter—took up not *where* matter is extracted from but that matter itself, to ask what part (if anything at all) of its material origins remain alive and participate in its postextraction existence. Liu Bolin's photograph, after which the section is named, shows a human figure stained by (what looks like) the coal piled behind him. The two chapters of this section equally attempted to see how exterranean materials become discernable and detectable, how we coexist with them as (in)visible affiliates. There could have been as many chapters as there are exterranean materials, ones on coal, oil, plastic, etc. Preferring an oblique to a direct approach, I selected limestone and salt for analysis. Chapter 6 read the early modern Normandy town of Caen by intercrossing human and geological timelines in order to bring out the role played by local limestone extraction in construction (of houses) and destruction (of churches), as well as the related human longings and anxieties. Chapter 7 switched from limestone to salt, to explore a wholly different exterranean matter but with whose efficacies (or *doings*) we humans live. In both chapters, there are moments when the matter's (former) connection to the Earth becomes visible and others where it remains unperceived. It emerged that no one thing governs (or can govern) this perceptibility. Generally, though, the proximity of (representations of) extraction (sites) to the exterranean matter turned out to be clearly one major factor, whether the nearness of Caen's limestone quarries to the town's buildings threatened by religious upheaval or the *saliera* on François Ier's table. In a word: We need reminders of exterraneanity.

As we progressed from section to section, from Terra, to *terra*, to exterranean matter, and even if (in early modern humanist terms—and as Montaigne proposed)[8] we might have seemed to be progressively swapping out cosmography for topography, we less switched scales than we sought to grasp, conceptually, exterranean activity at different moments of a process (Terra, before extraction; *terra*, during extraction; exterranean matter, after extraction), in order to perceive a sense of material and immaterial continuity. Perhaps we might say that, when we think exterraneanly, *there is only topography*. In Latourian terms: "connectivity, yes; scale, no."[9] And in Mortonian terms, each section, each conceptual grasping, has been something of a TARDIS, bigger on the inside than on the outside.[10] Exterranean matter is not some *smaller* thing than Terra; it just is (or was) *part of* it. As Latour puts it in *Où atterrir?* (*Where to Land?*), published as the writing of *Exterranean* drew to a close, the terrestrial (*terrestre*) is also global/worldly (*mondial*) in that it adheres to no frontier and defies all identitary boundaries.[11] So, too, for the exterranean. The exterranean has no politics of its own, and it belongs outright to no one scientific or nonscienfitic discipline—rather, it acquires from Terra, *terra*, and exterranean matters political and disciplinary affiliations, just as we Earthlings do.

Over the course of this book, each chapter took a hermeneutic wager:

1. Granting standing to Terra but not quite knowing who/what Terra is/does.
2. Seeking out Terra's *res* in poetry's *verba*, refusing to fall under mythology's spell—by taking mythology seriously.
3. Reading Terra for global geographies *and* for surface depths.
4. Reading a miner's bible *despite* its obvious emphasis on "how to," that is, τέχνη.
5. Understanding mining spirits as material, not cultural, phenomena.
6. Interweaving human and geological timelines to access limestone.
7. Giving to salt the full agency of its doings.

From such wagers resulted seven hypothetical investigations and seven final conclusions about the exterranean:

1. Extraction *ex Terra* locks whole and part together not across scales (big whole/small part) but as interconnected systems.
2. Proposed genealogies for exterranean matter can carry vibrancy across scales but can also efface material origins.
3. The exterranean can, despite the danger in seeing globes only from afar, be thought globally in order to allow for anthropological difference.
4. How we understand exterranean activity—instrumentalize Terra, give Terra a voice, or silence and ignore Terra—is ultimately a choice.
5. Do not let modernity explain away the demons that preside over exterranean activity.

6. Geomedia such as limestone remind us that humans have long been a geological force and also that we have long struggled with realizing it. Thus: *See the stone into which the image is sculpted.*

7. Exterranean agency is observable—but matter must be puzzled over.

The seven endzones leave us in the impersonal and horrific in-between that is anchored on one side to the "world-for-us" and on the other to the "world-for-itself" of which Eugene Thacker speaks in his *In the Dust of This Planet.*[12] Or rather—to avoid mismatching histories—those endzones leave us between the Terra-for-us and the Terra-for-itself. And they leave us between Terra-as-mass and Terra-as-living-system. In the Anthropocene, thinkers are calling on us to "think big" and to have a "sense of planet." This book has not argued against that—at least, not exactly. It has tried to listen to humanist voices that, sometimes despite themselves, articulate the exterranean as a connecting principle, as a coupler—perhaps, we might say, as a new and wholly material *religio*. To think the exterranean is to shuttle back and forth between instrumentalism and animism, to avoid choosing between seeing matter and seeing the material systems in which matter *is*. If the opposite of addiction is not sobriety but connection, then this book has been, more than anything, a moment in rehab whose goal was and is to wean us from our hedonistic and extractivist isolationism vis-à-vis Planet Earth as well as from our blindness to all the "bits" of Earth that are not just under us but above and around us and even in our bodies.[13] It has been an attempt to wake up from the dream according to which we humans can, guilt-free, ignorant of each other, and glancing at the happy smiling birds, *take-stuff-from-the-Earth* just like the singing miners in *Snow White and the Seven Dwarfs* (1937) who live on a delightful, joyous, and resource-infinite planet that simply does not exist.

ACKNOWLEDGMENTS

Between my earlier work on early modern space and this current project occurred what I casually call my "3D-turn." Making that turn opened up numerous interdisciplinary dialogues and brought me into contact with countless smart and generous scholars who offered both support and productive criticism along the way. In chronological order, I should like especially to thank: Susanne Wofford, a generous colleague and joyous organizer of the NYU Renaissance Salon, for offering this project its first public venue in February 2015; everyone at the UCSB Interdisciplinary Humanities Center's May 2015 conference on "Approaching the Anthropocene," especially Susan Derwin, Timothy Morton, and Volker M. Welter; David P. Laguardia for inviting me to speak on parts of this project at a Guthrie Workshop on "The Spaces of the French Renaissance" at Dartmouth College, also in May 2015, where many colleagues offered guidance, especially Leah Chang, Katie Chenoweth, Tom Conley, Andrea Frisch, and Dorothea Heitsch; Charles-Louis Morand-Métivier, who invited me to speak about this work at the University of Vermont in September 2015; Michael Meere, Nadja Aksamija, and the members of the Wesleyan Renaissance Seminar, for inviting me to talk in October 2015; Tom Conley and Sanam Nader-Esfahani for hosting me on "New World Mining in the Humanist Anthropocene" at the Renaissance Seminar at Harvard University's Mahindra Humanities Center in October 2015; Rebecca Weaver-Hightower, for welcoming me onto a panel about mining at the MLA in Austin, Texas, in January 2016, and for the vibrant dialogue over cocktails, along with fellow panelists Jennifer Blair and Pedro Garcia-Carol; Tom Conley and Katharina Piechocki, for putting this project around the table at the Radcliffe Seminar on "Cartography and Spatial Thinking from Humanism to the Humanities" in March 2016; Emily Apter for a lively and productive discussion of Chapter 6 in her graduate theory seminar at NYU in March 2016; Tom Conley and Mireille

Huchon for providing a Franco-American venue for the "Anthropocène à la Renaissance" at the Amis de l'Atelier workshop at Harvard University, also in March 2016; Emily Thompson and Colette Winn for allowing my stony stories into a conference on early modern narrative at Washington University in St. Louis, in April 2016; Frank Lestringant for inviting me to present a version of Chapter 1 at the Université de Paris–Sorbonne in February 2017; Timothy Chesters and Raphaële Garrod for welcoming me to Cambridge University's Early Modern French Seminar in March 2017, for a discussion of Chapter 3; and Frédérique Aït-Touati, Emanuele Coccia, and Philippe Quesne for a fabulous interdisciplinary exploration of Part II in an EHESS seminar on "La connaissance sensible" in March 2017. Many others—too many to name—offered help along the way. I should like to offer special thanks to Tom Conley, whose call to put "French Studies on the Map," not to mention his unstinting support and friendship over the years, have been essential in bringing me to a place where I could write the present book, and to Louisa Mackenzie, whose own work in rethinking the early modern in light of current theories and problems has been and remains a beacon. New York University, where I arrived in September 2014, quickly showed itself to be a fabulously exciting and collegial place to think and to write. All my departmental colleagues have offered intellectual and human support, for which I am immensely grateful, especially Emily Apter, Benoît Bolduc, and Sarah Kay. Some of this book was written over several summers at NYU Paris while running a dissertation writing workshop. Thank you to Benoît Bolduc, Jair Kessel, Nathalie Gicquel, as well as to several cohorts of graduate students who made my time so productive and enjoyable. The rest of the writing and revision happened during a year of sabbatical leave (2016–2017) made possible by the generous support of the American Council of Learned Societies (ACLS) and New York University's Office of the Dean for Humanities. My final thanks are to my family, for always being there, and to my breathtakingly smart wife, Penelope Meyers Usher, my intellectual and everything partner each and every day.

NOTES

INCIPIT: FROM SUB- TO EXTERRANEAN

1. Chapter epigraph sources: Michel Serres, *Le contrat naturel* (Paris: Flammarion, 1992), 16; *The Natural Contract*, trans. Elizabeth MacArthur and William Paulson (Ann Arbor: University of Michigan Press, 1995), 3. In French: ". . . les choses placées jadis là comme décor autour des représentations ordinaires, tout cela, qui n'intéressa jamais personne, brutalement, sans crier gare, se met désormais en travers de nos manigances." Jean Céard, *La nature et les prodiges* (Geneva: Droz, 1996), xii. Translation mine. In French: "L'homme du XVIᵉ siècle n'observe pas les choses d'un œil froid, d'un regard détaché; il lui semble avoir partie liée avec elles: elles sont des voix avec lesquelles toute enquête, toute relation est un véritable dialogue."

2. The concentration of CO_2 on any given day can be seen at https://www.co2.earth/. On the day I review these words, March 8, 2017, the concentration is 406.42 ppm. It first passed the milestone 400 ppm threshold in 2016. Brian Kahn, "Earth's CO_2 Passed the 400ppm Threshold—Maybe Permanently," *Scientific American*, September 27, 2016, https://www.scientificamerican.com/article/earth-s-co2-passes-the-400-ppm-threshold-maybe-permanently/.

3. Much has been written of late on these topics. On visions of the planet from the 1960s to the present day, see Ursula Heise, *Sense of Place and Sense of Planet* (Oxford: Oxford University Press, 2008), chap. 1, esp. 28–47. More specifically on the relationship between *Earthrise* and the rise of 1960s environmentalism, see Robert Poole, *Earthrise: How Man First Saw the Earth* (New Haven, CT: Yale University Press, 2008), chap. 8 ("From Spaceship Earth to Mother Earth"). See also, in the Object Lessons series, Jeffrey Jerome Cohen and Linda T. Elkins-Tanton, *Earth* (New York: Bloomsbury, 2017). For a certain history of this totalized view, see Ayesha Ramachandran, *The Worldmakers: Global Imagining in Early Modern Europe* (Chicago: University of Chicago Press, 2015); and my review thereof in *Imago Mundi* 69, no. 1 (2017): 132–33.

4. I develop these points further along, but the key reference is Bruno Latour, "L'Anthropocène ou la destruction (de l'image) du globe," in *Face à Gaïa. Huit conférences sur le*

nouveau régime climatique (Paris: La Découverte, 2016): 147–91. Something of this turn to soil and messy matter is captured—although with quite different means and ends—by the chapters of Jeffrey Jerome Cohen and Lowell Duckert, eds., *Elemental Ecocriticism: Thinking with Earth, Air, Water, and Fire* (Minneapolis: University of Minnesota Press, 2015).

5. Latour, "L'Anthropocène ou la destruction (de l'image) du globe," 179–80, 187. For further explication, see also his "Why Gaia Is Not the Globe—and Why Our Future Depends on Not Confusing the Two," lecture delivered at the Faculty of Arts, Aarhus University, Denmark, June 15, 2016, viewable on YouTube, https://goo.gl/6c9cNa.

6. Bruno Latour, "Some Advantages of the Notion of 'Critical Zone' for Geopolitics," *Procedia Earth and Planetary Science* 10 (2014): 3–6, 4. Latour further unpacks the concept in "The Notion of 'Critical Zones' and the Redefinition of Territories," lecture delivered at the Fondation de l'Écologie Politique, November 20, 2015, https://www.youtube.com/watch?v=QD3B5DR-x2E.

7. Bruno Latour, "Life among Conceptual Characters," *New Literary History* 47, nos. 2–3 (2016): 473.

8. Latour, "Life among Conceptual Characters," 472. For a further rapprochement of Latour's latest turn to "critical zones" and the early modern period, see Phillip John Usher, "The Revenge of the Mines: Earth-from-Nowhere versus Surfaces-with-Depths," in *Early Modern Visions of Space*, ed. Jeremie Korta and Dorothea Heitsch (Chapel Hill: University of North Carolina Press, forthcoming).

9. Bruno Latour, *Où atterrir?* (Paris: La Découverte, 2017), 87.

10. Latour develops the call for descriptive approaches in Latour, *Où atterrir?*, 119–25. See also Usher, "The Revenge of the Mines." *Exterranean*, taking up the terrestrial and the exterranean, thus also stands in opposition to the globe-focused history of the early modern offered by Ramachandran, *The Worldmakers*.

11. The usage of *Terra* and *Tellus* has evolved over time. Servius differentiates between *Terra*, the earth considered as one of the four elements (with fire, water, and air), and *Tellus*, the deity who guards the Earth, i.e., Gaia. Servius, *In Vergilii carmina comentarii* (Leipzig: B. G. Teubner, 1881), 171, http://www.perseus.tufts.edu/hopper/text?doc=Serv.+A.+1.171&fromdoc=Perseus%3Atext%3A1999.02.0053. See also the entry on "Tellus" in *The Oxford Classical Dictionary* (Oxford University Press, 1996), 1480.

12. *OED*, "mine": "Anglo-Norman and Middle French miner, myner, French miner (c1200 in Old French in sense 'to dig under land, a rock, a wall, etc., so as to make it collapse', c1340 in figurative sense of 'to destroy, wear down', 15th cent. in sense 'to hollow out by erosion', c1550 in sense 'to extract from a mine', 1680 in sense 6), probably < mine n. Compare Old Occitan minar (14th–15th cent.), Italian minare (a1348), Spanish minar (1495), Portuguese minar (16th cent.), post-classical Latin minare (13th cent.), mineare (13th cent. in British sources); also Dutch mijnen , German minieren (early 16th cent.), Norwegian mine to blast, blow up, Swedish mina (second half of the 17th cent.), Danish mine."

13. The expression "ex terra" (i.e., "from the earth/Earth") appears, of course, in numerous Latin texts, both classical and postclassical. Augustine talks of things com-

ing "ex terra" in a discussion of the ultimate earthly origins of trees and animals in Augustine, *Œuvres, De Genesi ad litteram* [Literary commentary on Genesis], 23.44 (Paris: Desclée de Brouwer, 1972), 48:437–39. Closer to the concerns and texts of *Exterranean*, as will become clear, is Georgius Agricola's *De natura eorum quae effluunt ex terra* (Basel: per Hieronymum Frobenium et Nic. Episcopium, 1546) (consulted at Columbia University *RBML* B508 Ag8). See also Agricola's use of "ex terra" in his *De animantibus subterraneis liber* (1549) in Agricola, *De re metallica traduit de l'édition originale latine de 1556* (Thionville: Gérard Hlopp, 1992), 504–5. My coining of the term *exterranean* departs, of course, from the Latin adjective *exterraneus*, from "ex alia terra," i.e., "which comes from another land/country" (see Lewis and Short).

14. Dipesh Chakrabarty, "Postcolonial Studies and the Challenge of Climate Change," *New Literary History* 43 (2012): 14.

15. By talking of untranslatability, I draw on Barbara Cassin, *Vocabulaire européen des philosophies: Dictionnaire des intraduisibles* (Paris: Le Seuil, 2004); *Dictionary of Untranslatables: A Philosophical Lexicon*, ed. Emily Apter, Jacques Lezra, and Michael Wood (Princeton, NJ: Princeton University Press, 2014). For a theoretical exploration of this project, see in particular Emily Apter, "Untranslatables: A World System," *New Literary History* 39, no. 3 (2008), 581–98; and Emily Apter, *Against World Literature: On the Politics of Untranslatability* (New York: Verso, 2013). "Untranslatability" does not mean that a term *cannot* be translated. It refers rather to the process by which a concept or an idea circulates between languages in an ongoing process of translation, mistranslation, and retranslation. For a longer untranslation of "Anthropocene," see Phillip John Usher, "Untranslating the Anthropocene," *Diacritics* 44, no. 3 (2016): 56–77.

16. Illustrations from Alphonse Bertillon, *Identification anthropométrique* (Paris: Ministère de l'Intérieur, 1885).

17. Paul J. Crutzen and Eugene F. Stoermer, "The 'Anthropocene,'" *Global Change Newsletter* 41 (May 2000): 17–18. All emphases mine.

18. Space does not permit an exhaustive rehearsal of the history of the term "humanism" here. What follows seeks to make some key points to differentiate early modern humanism from later incarnations and to assert what the *homo* of early modern humanism might be. Fuller histories, on which these remarks draw, include Christopher C. Celenza, "Humanism," in *The Classical Tradition*, ed. Anthony Grafton, Glenn W. Most, and Salvatore Settis (Cambridge, MA: Harvard University Press, 2010), 462–67; Vito R. Giustiniani, "Homo, Humanus, and the Meanings of 'Humanism,'" *Journal of the History of Ideas* 46 (1985): 167–95; Anthony Grafton and Lisa Jardine, *From Humanism to the Humanities: The Institutionalizing of the Liberal Arts in Fifteenth- and Sixteenth-Century Europe* (Cambridge, MA: Harvard University Press, 1986); Paul Oskar Kristeller, "The Humanist Movement," in *Renaissance Thought: The Classic, Scholastic, and Humanist Strains* (New York: Harper and Row, 1961), 3–23; and "Humanism," in *The Cambridge History of Renaissance Philosophy*, ed. C. B. Schmitt, Quentin Skinner, Eckhard Kessler, and Jill Kraye (Cambridge: Cambridge University Press, 2008), 113–37; Nicholas Mann, "The Origins of Humanism," in *The Cambridge Companion to Renaissance Humanism*, ed. Jill Kraye (Cambridge: Cambridge University Press, 2006), 1–19. Also particularly

pertinent in this volume are the articles by Jill Kraye, "Philologists and Philosophers" (142–60); and Anthony Grafton, "The New Science and the Traditions of Humanism" (203–23).

19. Kenneth Gouwens, "What Posthumanism Isn't," in *Renaissance Posthumanism*, ed. Joseph Campana and Scott Maisano (New York: Fordham University Press, 2016), 38.

20. Cary Wolfe, *What Is Posthumanism?* (Minneapolis: University of Minnesota Press, 2010), xi, xvi.

21. Gouwens, "What Posthumanism Isn't," 39.

22. Rosi Braidotti, *The Posthuman* (Cambridge: Polity, 2013).

23. On these questions, including the history of how Pico's text came to be seen as a celebration of human exceptionalism rather than a text about the extinction of self before the divine, see the concise and convincing section titled "Renaissance Conceptions of Human Dignity and Misery," in Gouwens, "What Posthumanism Isn't," 44–49.

24. Michel de Montaigne, *Essais*, ed. Pierre Villey and V. L. Saulnier (Paris: PUF, 2004), 452; *Essays*, trans. Donald Frame (New York: Everyman's Library, 2003), 330. Translation modified. Montaigne continues as follows, to situate humanity at the lowest level of the *scala naturae*: "Elle se sent et se void logée icy, parmy la bourbe et le fient du monde, attachée et clouée à la pire, plus morte et croupie partie de l'univers, au dernier estage du logis et le plus esloigné de la voute celeste, avec les animaux de la pire condition des trois; et se va plantant par imagination au dessus du cercle de la Lune et ramenant le ciel soubs ses pieds" ("[this creature] sees itself lodged here, amid the mire and dung of the world, nailed and riveted to the worst, the deadest, and the most stagnant part of the universe, on the lowest story of the house and the farthest from the vault of heaven, with the animals of the worst condition of the three; and in its imagination it goes planting itself above the circle of the moon, and bringing the sky down beneath his feet"). Montaigne, *Essais*, 452; trans. 330–31. Translation modified.

25. This example is taken from *Ephémérides du Citoyen* 1, no. 16 (December 27, 1765), 246–47, quoted by Giustiani, "Homo, Humanus, and the Meanings of 'Humanism,'" 175n38. In French: "L'amour général de l'humanité . . . vertu qui n'a point de nom parmi nous et que nous oserions appeler 'humanisme,' puisqu'enfin il est temps de créer un mot pour une chose si belle et nécessaire." The author's use of the verb *oser* (to risk, to dare) surely captures the term's perceived novelty.

26. Giustiani, "Homo, Humanus, and the Meanings of 'Humanism,'" 178.

27. See Nishitani Osamu, "Anthropos and Humanitas: Two Western Concepts of 'Human Being,'" trans. Trent Maxey, in *Translation, Biopolitics, Colonial Difference*, ed. Naoki Sakai and Jon Solomon (Hong Kong: Hong Kong University Press, 2006), 259–273; and Cassin, *Dictionary of Untranslatables*, "Humanity" (461) and esp. "Menschheit/Humanität" (650–653).

28. Robert E. Proctor, *Defining the Humanities* (Bloomington: Indiana University Press, 1998), 15.

29. As François Prost writes, in an *encadré* within the entry for *Menschheit* (Cassin, *Dictionary of Untranslatables*, 651), *humanitas* "establishes itself as a set of characteristics

that supposedly define what a civilized man is, as opposed to what he is not, and from which follow certain duties he has to observe in relation to himself, and to his fellow humans."

30. Terence, *The Self-Tormentor*, Loeb Classical Library (Cambridge, MA: Harvard University Press, 2014), l. 77.

31. The demarcation between Latin and Greek humans is further clarified by Nishitani ("Anthropos and Humanitas," 260), who summarizes concisely as follows: Whereas the *anthropos* "cannot escape the status of being the object" of knowledge, the *homo* "is never defined from without"—rather, the homo "expresses itself as the subject" of knowledge.

32. On Georg Voigt's key role in defining the term, see Paul F. Grendler, "Georg Voigt: Historian of Humanism," in *Humanism and Creativity in the Renaissance*, ed. Christopher S. Celenza and Kenneth Gouwens (Leiden: Brill, 2006), 293–325.

33. Giustiani, "Homo, Humanus, and the Meanings of 'Humanism,'" 171. Giustiani underscores that the term, at this point in time, "placed no particular emphasis on all the values entailed by the Latin term *humanus* in its broadest sense." Moreover, the sense of *humanae litterae* was also restricted, including neither law, medicine, nor the Bible (*litterae divinae*) but only "profane literature and grammar."

34. Bruno Latour, *Nous n'avons jamais été modernes* (Paris: La Découverte, 1997), 16, 44; *We Have Never Been Modern*, trans. Catherine Porter (Cambridge, MA: Harvard University Press, 1993), 7, 28.

35. Louisa Mackenzie, "It's a Queer Thing: Early Modern French Ecocriticism," *French Literature Studies* 39 (2012): 21.

36. Mackenzie, "It's a Queer Thing," 21–22; quoting Jeffrey Theis, *Writing the Forest in Early Modern England: A Sylvan Pastoral* Nation (Pittsburgh, PA: Duquesne University Press 2009), 43.

37. The best introductions to this body of writings, on which I draw here, are provided by Pamela O. Long, *Openness, Secrecy, Authorship: Technical Arts and the Culture of Knowledge from Antiquity to the Renaissance* (Baltimore, MD: Johns Hopkins University Press, 2001), 175–209; and Marie-Claude Déprez-Masson, *Technique, mot et image: Le De re metallica d'Agricola* (Turnhout: Brepols, 2006), 21–30.

38. A fuller pre–early modern overview is provided by Déprez-Masson, *Technique, mot et image*, 11–21. Robert Halleux explores how "la nouvelle structuration de savoirs techniques [new structuring of technical knowledge]" about mining evolves in the early modern period in his article "Le mineur et l'alchimiste. La systématisation des savoirs aléatoires au XVIᵉ siècle," in *Hasard et Providence au XIVᵉ–XVIIᵉ siècles*, ed. Marie-Luce Demonet (Tours: Centre d'Études Supérieures de la Renaissance, 2007), 1–12.

39. Long, *Openness, Secrecy, Authorship*, 177. See the modern English edition: Ulrich von Kalbe, *Bergbüchlein. The Little Book of Ores*, trans. Anneliese Sisco, ed. Cyril Stanley Smith and Mary Ross (England: Oxshott, 2013).

40. An English translation is included in *Bergwerk und Probierbüchlein*, trans. Anneliese Sisco, ed. Cyril Stanley Smith (New York: American Institute of Mining and Metallurgical Engineers, 1949), 77–188.

41. Vannoccio Biringuccio, *The Pirotechnia*, trans. and ed. Cyril Stanley Smith and Martha Teach Gnudi (New York: Dover, 1990).

42. Long, *Openness, Secrecy, Authorship*, 182. On this extraordinary book, see Wolfgang Lefèvre, "Picturing the World of Mining in the Renaissance: The *Schwazer Bergbuch* (1556)" (Berlin: Max-Planck-Institut für Wissenschaftsgeschichte, 2010), http://hdl.handle.net/11858/00-001M-0000-002A-7F8F-B. There are seven extant manuscripts of the *Schwazer Bergbuch* (Lefèvre, "Picturing," 112). For a modern edition and translation, see Heinrich Winkelmann, ed., *Schwazer Bergbuch: Codex Vindobonensis 10.852* (Essen: Verl. Glückauf, 1988).

43. Samuel Zimmermann, *Probierbüch: Auff alle Metall Müntz / Ertz / und berckwerck / Dessgleichen auff Edel Gestain / Perlen / Corallen / und andern dingen mehr* (Augsburg: Michael Manger, 1573). See Long, *Openness, Secrecy, Authorship*, 191.

44. Modern editions in German and English are available: Lazarus Ercker, *Beschreibung der Allerfürnemsten mineralischen ertz und Berckwerks arten von Jahre 1580*, ed. Paul Reinhard Beirlein and Afred Lange (Berlin: Akademie-Verlag, 1960); *Treatise on Ores and Assaying*, trans. Anneliese Grünhaldt Sisco and Cyril Stanley Smith (Chicago: University of Chicago Press, 1951). For a recent discussion of Ercker, see Pamela O. Long, *Openness, Secrecy, Authorship*, 188–91.

45. Ciriacus Schreittmann, *Probierbüchlin / Frembde und subtile Künst / vormals im Truck nie gesehen* (Frankfurt: Christian Egenolffs Erben, 1580). On Schreittmann, see Long, *Openness, Secrecy, Authorship*, 191.

46. Modestin Fachs, *Probier Büchlein / Darinne Gründlicher bericht vormeldet / wie man alle Metall / und derselben zugehörenden Metallischen Ertzen und getöchten ein jedes auff seine eigenschafft und Metall recht Probieren sol* (Leipzig: Zacharias Berwald, 1595). On Fachs, see Long, *Openness, Secrecy, Authorship*, 191.

47. The formula describing the *De re metallica* as the "miner's Bible" comes from Gray A. Brechin, *Imperial San Francisco: Urban Power, Earthly Ruin* (Berkeley: University of California Press, 1999), 25.

48. The chapters of this book study texts written in a number of languages, especially Latin and French, but also (to a lesser degree) German, Italian, and Spanish. It is beyond the scope of the present work to offer a full reflection on the stakes of texts having been written in one language or another. Such a study, building inter alia on Michael Cronin, *Eco-Translation: Translation and Ecology in the Age of the Anthropocene* (London: Routledge, 2017); Ursula Heise, "Comparative Ecocriticism in the Anthropocene," *Komparatistik* (May 2014): 19–33; and especially Cassin's notion of untranslatability, would be valuable indeed, to bring out even more clearly how the exterranean, because of its inherent materiality, unable to be grasped in its totality, is also always perceived *in a given language* although always too in *a given language* that must always be untranslated in Cassin's sense—as we shall see in Part I, where Greek *Gaia* brushes up against Latin *Terra* and French *terre*.

49. "These books transformed mining from a relatively low-status occupation into a learned subject with ancient precedents, a contribution to humanist learning." Long, *Openness, Secrecy, Authorship*, 177.

50. "Agricola portrays the interlocutor Bermannus as a model who combines direct observation and experience with knowledge of ancient texts." Long, *Openness, Secrecy, Authorship*, 185.

51. Latin text quoted from Georgius Agricola, *Bermannus (le mineur): Un dialogue sur les mines*, ed. Robert Halleux and Albert Yans (Paris: Les Belles Lettres, 1990), 10. Translation mine.

52. Agricola, *Bermannus*, 17.

53. In its opening sentences, Agricola (*Bermannus*, 9) notably bemoans how the names of things have become obscured and deformed as older Greek and Latin names were buried under vernacular terms.

54. Agricola, *Bermannus*, 35.

55. Agricola, *Bermannus*, 75.

56. Agricola, *Bermannus*, 81.

57. Robert Halleux ("Le mineur et l'alchimiste," 4) concludes his study of the status of mining and alchemy in this period by asserting that "La naissance du capitalisme moderne implique un changement d'échelle et la mobilisation de moyens humains et matériels plus importants. Les investisseurs ne peuvent se satisfaire de l'aléatoire, ils exigent que la part du risque soit réduite" ("The birth of modern capitalism implies a change in scale and the mobilization of greater human and material means. Investors could not accept the idea of chance; rather, they demanded that risk be reduced to a minimum").

58. Long, *Openness, Secrecy, Authorship*, 186.

59. Simon L. Lewis and Mark A. Maslin, "Defining the Anthropocene," *Nature* 519 (March 12, 2015): 171–80. It is noteworthy that Lewis and Maslin translate *orbis* as "world" (175) rather than "globe," a choice that has far-reaching consequences given the Heideggerian weight of the former term. See Pascal David's entry on "Welt" in Cassin, *Dictionary of Untranslatables*, 1217–24.

60. The idea of the Columbian Exchange regarding the massive transfer of plants, animals, diseases, culture, technology, and peoples between the Old and New Worlds was put forward by Alfred W. Crosby, *The Columbian Exchange: Biological and Cultural Consequences of 1492* (Westport, CT: Praeger, 2003). Although unnoticed at its first publication in 1972, Crosby's book has now come to be recognized as "a founding text in the field of environmental history." Megan Gambino, "Alfred W. Crosby on the Columbian Exchange," *Smithsonian.com*, October 2, 2011, http://www.smithsonianmag.com/history/alfred-w-crosby-on-the-columbian-exchange-98116477/?all. Crosby, before the advent of the term "Anthropocene," explored explicitly some of the specifically ecological aspects of early modern colonialism in his *Ecological Imperialism: The Biological Expansion of Europe, 900–1900* (Cambridge: Cambridge University Press, 1986). There is unsurprisingly some uncertainty as to the specific number of people killed in the aftermath of 1492. Recent estimates quoted by Lewis and Maslin ("Defining," 175) suggest a fall from 54–61 million to about 6 million, i.e., a fall of between 48 and 55 million. Lewis and Maslin (176–77) also suggest another possible starting date for the Anthropocene, 1964, based on the global peak of ^{14}C from radioactive fallout.

61. Lewis and Maslin, "Defining," 175. A GSSP is an internationally recognized reference point on a stratigraphic section that serves to define the lower boundary of a stage on the geologic timescale. Official efforts to define GSSPs are managed by the International Commission on Stratigraphy.

62. Lewis and Maslin, "Defining," 175.

63. Dana Luciano, "The Inhuman Anthropocene," *Avidly—The Los Angeles Review of Books,* March 22, 2015, http://avidly.lareviewofbooks.org/2015/03/22/the-inhuman-anthropocene.

64. Emphasis mine.

65. Wynter argues as follows at the start of her chapter: "[The] basis of my proposed human view of 1492 . . . is that both the undoubted 'glorious achievement' of the processes that led up to Columbus's realization of his long dreamed-of voyage and the equally undoubted horrors that were inflicted by the Spanish conquistadores and settlers upon the indigenous peoples of the Caribbean and the Americas, as well as upon the African-descended Middle Passages and substitute slave labor force, are to be seen as the effects of Western Europe's epochal shift. That shift—out of the primarily supernaturally guaranteed modes of 'subjective understanding' (and, therefore, of their correlated symbolic-representational and ethico-behavioral systems) that had been common to all human cultures and their millennial traditional 'forms of life'—was a product of the intellectual revolution of humanism. Elaborated by humanists as well as by monarchical jurists and theologians, this revolution opened the way toward an increasingly secularized, that is, degodded, mode of 'subjective understanding.'" Sylvia Wynter, "1492: A New World View," in *Race, Discourse, and the Origin of the Americas: A New World View,* ed. Vera Lawrence Hyatt and Rex Nettleford (Washington, DC: Smithsonian Institution Press, 1995), 13.

66. Without referencing the Orbis hypothesis or even the events that followed 1492, in a recent interview with Natasha Lennard published in Simon Critchley's series *The Stone* at the *New York Times,* Cary Wolfe made a similar amalgam to Luciano's in his attempt to answer the question "Is Humanism Really Humane?" *New York Times,* January 9, 2017, https://nyti.ms/2jjVBGm. See also my reaction to the interview: Phillip John Usher, "Wrong Question: Is Humanism Humane?" *The Humanist Anthropocene,* http://thehumanistanthropocene.weebly.com/blog/wrong-question-is-humanism-humane, January 10, 2017.

67. Steve Mentz, "Enter Anthropocene, c. 1610," *Arcade: Literature, the Humanities, & the World,* December 1, 2015, http://arcade.stanford.edu/blogs/enter-anthropocene-c1610. An earlier version of this article, which does not mention the Capitalocene, appeared under the same title in *Glasgow Review of Books,* September 27, 2015, https://glasgowreviewofbooks.com/2015/09/27/enter-anthropocene-c-1610. In fact, Mentz's response even perhaps emphasizes the nonhuman so far as to downplay the inhumanity of the post-1492 genocide, as when he writes: "Reconsidering the 1610 Anthropocene through both capitalist expansions and more-than-human collisions helps emphasize that the core story, the story that still needs telling and that meaningfully precedes the

supposed modernity of the past half-millennium, concerns the production of hybrids through the collision of Unlike Worlds."

68. The messiness and swirl of history opened by the 1610 date bring the Anthropocene closer to Mentz's own concepts (a "cene salad"), including the Homogenocene, the Thalassocene, and the Naufragocene, which he develops in *Shipwreck Modernity: Ecologies of Globalization, 1550–1719* (Minneapolis: University of Minnesota Press, 2015), xii–xxiii. Mentz talks of "beheading Anthropos" on p. xii.

69. More scholars in the humanities have responded to the Orbis hypothesis than can be discussed here, including Nancy J. Jacobs, Danielle Johnstone, and Christopher S. Kelly, "The Anthropocene from Below," in *World Histories from Below: Disruption and Dissent, 1750 to the Present*, ed. Antoinette Burton and Tony Ballantyne (New York: Bloomsbury, 2016), 197–230, esp. 217; Jared Hickman, *Black Prometheus: Race and Radicalism in the Age of Atlantic Slavery* (New York: Oxford University Press, 2016), chap. 1; and Sharae Deckard, "Latin America in the World-Ecology: Origins and Crisis," in *Ecological Crisis and Cultural Representation in Latin America*, ed. Mark Anderson and Zélia M. Bora (New York: Lexington, 2016), 3–20, esp. 6.

70. Especially important are Michel Serres, *Le contrat naturel* (*The Natural Contract*); and Bruno Latour, *Nous n'avons jamais été modernes* (*We Have Never Been Modern*). On the tradition of *éco-pensée*, see Stephanie Posthumus, "État des lieux de la pensée écocritique française," *Eco-zone. European Journal of Literature, Culture, Environment* 1, no. 1 (2010): 148–54; Stephanie Posthumus, "Translating Ecocriticism: Dialoguing with Michel Serres," *Reconstruction* 7, no. 2 (2007), http://reconstruction.eserver .org/Issues/072/posthumus.shtml. It is important to note here that Serres's notion of a "natural contract" and Timothy Morton's rejection of nature and sculpting of the concept of "the mesh" are less incompatible than first meets the eye: Serres and Morton use the word "nature" in different ways, but both seek out ways of mapping interconnectedness and entanglement.

71. Timothy Morton, *Ecology without Nature* (Cambridge, MA: Harvard University Press, 2007); *The Ecological Thought* (Cambridge, MA: Harvard University Press, 2010); *Hyperobjects: Philosophy and Ecology after the End of the World* (Minneapolis: University of Minnesota Press, 2013); and *Dark Ecology: For a Logic of Future Coexistence* (New York: Columbia University Press, 2016). Jane Bennett, *Vibrant Matter: A Political Ecology of Things* (Durham, NC: Duke University Press, 2010). Quentin Meillassoux, *Après la finitude: Essai sur la nécessité de la contingence* (Paris: Seuil, 2012); *After Finitude: An Essay on the Necessity of Contingency*, trans. Ray Brassier (New York: Continuum, 2010). Graham Harman, *Tool-Being: Heidegger and the Metaphysics of Objects* (Chicago: Open Court, 2002); *Quentin Meillassoux: Philosophy in the Making* (Edinburgh: Edinburgh University Press, 2011). Levi Bryant, *The Democracy of Objects* (Ann Arbor: Open Humanities Press, 2011). Tristan Garcia, *Forme et objet: Un traité des choses* (Paris: PUF, 2011).

72. My aim here is not to be cavalier but rather (a) to respect the complexity of these various thinkers by not providing oversimplified summaries and (b) to not overly slow down my own writing, which is always a risk: One reviewer of Cohen's *Stone* notably

commented that doing "his due theoretical diligence [in *Stone*, 39–47] dull[ed] [the] writing." Paul A. Harris, review of *Stone*, *SubStance* 45, no. 2 (2016): 186.

73. Bruce Boehrer, *Environmental Degradation in Jacobean Drama* (Cambridge: Cambridge University Press, 2013); Todd Borlik, *Ecocriticism and Early Modern English Literature: Green Pastures* (New York: Routledge, 2011); Gabriel Egan, *Green Shakespeare: From Ecopolitics to Ecocriticism* (New York: Routledge, 2006); Thomas Hallock, Ivo Kamps, and Karen Raber, eds., *Early Modern Ecostudies: From the Florentine Codex to Shakespeare* (New York: Palgrave Macmillan, 2008); Ken Hiltner, *Milton and Ecology* (Cambridge: Cambridge University Press, 2003); Ken Hiltner, ed., *Renaissance Ecology: Imagining Eden in Milton's England* (Pittsburgh, PA: Duquesne University Press, 2008); Ken Hiltner, *What Else Is Pastoral? Renaissance Literature and the Environment* (Ithaca, NY: Cornell University Press, 2011); Steve Mentz, *At the Bottom of Shakespeare's Ocean* (London: Continuum, 2009); Steve Mentz, *Shipwreck Modernity: Ecologies of Globalization, 1550–1719* (Minneapolis: University of Minnesota Press, 2015); Vin Nardizzi, *Wooden Os: Shakespeare's Theatres and England's Trees* (Toronto: University of Toronto Press, 2013); Robert Watson, *Back to Nature: The Green and the Real in the Late Renaissance* (Philadelphia: University of Pennsylvania Press, 2006); Tiffany Werth, "A Heart of Stone: The Ungodly in Early Modern England," in *The Indistinct Human in Renaissance Literature*, ed. Jean Feerick and Vin Nardizzi (New York: Palgrave, 2012), 181–204; Tiffany Werth, "Loving London Stone," *Upstart: A Journal of Renaissance English Studies* (February 14, 2014), http://www.clemson.edu/upstart/Essays/london_stone/london_stone.xhtml; Lynne Bruckner and Dan Brayton, eds., *Ecocritical Shakespeare* (Burlington, VT: Ashgate, 2011).

74. For up-to-date information on this series, edited by Gilliam Overling, Heide Estes, Philip Slavin, and Steve Mentz, see http://en.aup.nl/series/environmental-humanities-in-pre-modern-cultures.

75. Mackenzie, "It's a Queer Thing," 20.

76. Karen Thornber, *Ecoambiguity: Environmental Crises and East Asian Literature* (Ann Arbor: University of Michigan Press, 2012). Thornber's volume was well reviewed and received honorable mentions for the Rene Wellek Prize of the ACLA (2013) and for the ASLE's Best Book-Length Monograph on Scholarly Ecocriticism Prize (2013).

77. Jeff Persels, ed., *The Environment*, special issue of *French Literature Studies* 39 (2012); Douglas L. Boudrieu and Marnie M. Sullivan, eds., *Ecocritical Approaches to Literature in French* (Lanham, MD: Lexington, 2015); Pauline Goul and Phillip John Usher, eds., *Early Modern Écologies* (Amsterdam: Amsterdam University Press, 2019); Pasquale Verdicchio, ed., *Ecocritical Approaches to Italian Culture and Literature: The Denatured Wild* (Lanham, MD: Lexington, 2016); Serenella Iovino, ed., *Ecocriticism and Italy: Ecology, Resistance, and Liberation* (New York: Bloomsbury, 2016); Chia-ju Chang and Scott Slovic, eds., *Ecocriticism in Taiwan: Identity, Environment, and the Arts* (Lanham, MD: Lexington, 2015). See also Karen Thornber, "Literature, Asia, and the Anthropocene: Possibilities for Asian Studies and the Environmental Humanities," *Journal of Asian Studies* 73, no. 4 (November 2014): 989–1000; Cheng Li, "Echoes from the Opposite Shore: Chinese Ecocritical Studies as a Transpacific Dialogue Delayed," *ISLE* 21, no. 4 (Autumn 2014):

821–43; John Charles Ryan, "Beyond East Asian Landscapes: A Survey of Ecocriticism in Thai Literary Studies," International Conference on Trends in Economics, Humanities and Management (ICTEHM'15), August 12–13, 2015, Pattaya (Thailand), http://icehm .org/upload/3350ED815006.pdf; Rayson K. Alex, "A Survey of the Phases of Indian Ecocriticism," *CLCWeb* 16, no. 4 (2014): http://docs.lib.purdue.edu/clcweb/vol16/iss4/9/.

78. Alain Suberchicot offers a comparative analysis of French, American, and Chinese perspectives in his *Littérature et environnement. Pour une écocritique comparée* (Paris: Champion, 2012). More recently, Ursula Heise has taken up the question of comparative literature's relationship with ecocrticism and ecology: "Comparative Literature and the Environmental Humanities," ACLA State of the Discipline Report (March 9, 2014), https://stateofthediscipline.acla.org/entry/comparative-literature-and-environmental -humanities; Heise, "Comparative Ecocriticism in the Anthropocene."

79. The most thorough treatment of this question is Cronin, *Eco-Translation*.

80. See https://thewire.mla.hcommons.org/ecocriticism-environmental-humanities/. In an online comment below the forum's self-definition, Beatriz Celaya, who teaches in the Spanish section of the Department of Romance Languages and Literatures at the University of Cincinnati, responded: "I am not in English Studies, should I not be interested in ecocriticism? I hope the mistake was involuntary."

81. Louisa Mackenzie's chapter "The Poet and the Environment: Naturalizing Conservative Nostalgia" takes up the poets Pierre de Ronsard and Jean-Antoine de Baïf to see how early modern French poetry engages with environmental history: see her *The Poetry of Place: Lyric, Landscape, and Ideology in Renaissance France* (Toronto: University of Toronto Press, 2011), 121–45. The diagnosis that Mackenzie ("It's a Queer Thing," 21) made in 2012 is still largely true today: "There is almost nothing on non-Anglophone early modernity."

82. Jeffrey Jerome Cohen, *Stone: An Ecology of the Inhuman* (Minneapolis: University of Minnesota Press, 2015).

83. My first public presentation of *Exterranean* dates from February 2015, as detailed at http://thehumanistanthropocene.weebly.com/blog/archives/02-2015; Cohen's book appeared on May 6, 2015.

84. Cohen (Stone, 10) explains in his introduction that, in order to plumb "the petric in the human and the anthropomorphic in the stone," he speaks "of the 'inhuman' to emphasize both difference ('in' as negative prefix) and intimacy ('in-' as indicator of estranged interiority)."

85. Cohen, *Stone*, 16, 40.

86. Montaigne, *Essais*, 657. My translation.

87. I think here of Louisa Mackenzie, who opposes "additive" interdisciplinarity and ecotheoretical approaches that cross boundaries.

88. For more on early modern and contemporary geographic scale, see Phillip John Usher, *L'aède et le géographe. Poésie épique et espace du monde à l'époque pré-moderne* (Paris: Classiques Garnier, 2018), esp. the Introduction and Conclusion.

89. It is, in fact, despite usage, not accurate to call the play simply Bruno Latour's. Bruno Latour authored with Chloé Latour and Frédérique Aït-Touati a first

play titled *Cosmocolosse*, which was followed by *Gaïa Global Circus*, written by Pierre Daubigny.

90. See Latour, "L'Anthropocène ou la destruction (de l'image) du globe," 147–91.

91. Sebastian Münster, *Cosmographiae universalis lib. VI* (Basel: apud Henrichum Petri, 1552), 11.

92. A gore is a flat section of a globe; in this context it refers to the printed flat paper panels that could be affixed to a sphere.

PART I. TERRA GLOBAL CIRCUS

1. Basic information about the play, including that quoted here, is available in the unsigned "A propos de Gaïa Global Circus (GGC). Réponse à quelques questions fréquentes (FAQ)," http://www.bruno-latour.fr/sites/default/files/downloads/FAQ%20GAIA GLOBALCIRCUS_0.pdf.

2. The text of *Cosmocolosse* (version 4, dated May 2011) is available at http://www .bruno-latour.fr/sites/default/files/downloads/GAIA-VERSION%2005-11-ssimage_1.pdf. An audio staging of the play can be listened to at https://www.franceculture.fr/emissions /latelier-fiction/cosmocolosse-de-bruno-latour-frederique-ait-touati-et-chloe-latour.

3. *Cosmocolosse*. My translation. French: "Elle enregistre beaucoup mieux l'incertitude où se trouvent tous les protagonistes sur qui parle et de quoi quand il s'agit de la terre."

4. Rania Ghosn and El Hadi Jazairy, "Gaïa Global Circus: A Climate Tragicomedy," in *Climates: Architecture and the Planetary Imaginary*, ed. James Graham, special issue of *Avery Review* (2016): 53.

5. Quoted from the teaser video made available by the Théâtre Dijon Bourgogne: https://www.youtube.com/watch?v=uMlfgMU6qG0, published September 21, 2016. My translation. French: "Maintenant je suis Gaïa et toi aussi, connard, tu es Gaïa."

1. TERRA HAS STANDING

1. Paulus Niavis, *Judicium Jovis*, in Johann Friedrich Klotzsch and Gottfried Immanuel Grundig, eds., *Sammlung vermischter Nachrichten zur Sächsischen Geschichte* (Chemnitz: Stobel, 1767), 51. Unless otherwise noted, all references are to this edition. Translations mine. The earliest edition is available as *Iudicium Iovis ad quod mortalis homo a terra tractus parricidii accusatus* (Leipzig: Martin Landsberg, 1495), Bayerische Staatsbibliothek, call number 4 Inc.s.a. 1334. Toward the end of my research, I discovered the existence of Frank Dawson Adam's manuscript English translation of the *Judicium Jovis*, now at the library of McGill University, catalogued under the title *This book tells of a trial conducted by Jupiter in the Vale of Beauty, to which a mortal man was brought by Earth for digging into Mt. Niveus, and many other mountains and arraigned on no less a charge than parricide* (c. 1930), call number N577j 1930za. Adam likely prepared that translation while working on his *The Birth and Development of the Geological Sciences* (Baltimore, MD: The Williams and Wilkins Company, 1938).

2. Niavis, *Judicium Jovis*, 51.

3. Niavis, *Judicium Jovis*, 51.

4. Niavis, *Judicium Jovis*, 52.

5. Niavis, *Judicium Jovis*, 53.

6. Niavis, *Judicium Jovis*, 53.

7. Niavis, *Judicium Jovis*, 56.

8. *Homeric Hymns. Homeric Apocrypha. Lives of Homer*, trans. Martin L. West, Loeb Classical Library (Cambridge, MA: Harvard University Press, 2003), 212–15.

9. *Homeric Hymns. Homeric Apocrypha. Lives of Homer*, 212–215.

10. *The Orphic Hymns*, trans. Apostolos N. Athanassakis and Benjamin M. Wolkow (Baltimore, MD: Johns Hopkins University Press, 2013), 24.

11. *The Orphic Hymns*, 24 (editor's note). See Timothy Morton, *Ecology without Nature: Rethinking Environmental Aesthetics* (Cambridge, MA: Harvard University Press, 2007).

12. I draw here on Peter Dronke, "Bernard Silvestris, Natura, and Personnification," *Journal of the Warburg and Courtauld Institutes* 43 (1980): 20.

13. For a detailed analysis of these codices, see John McEnerney, "*Precatio terrae* and *Precatio omnium gherbarum*," *Rheinisches Museum für Philologie* 126 (1983): 175–87.

14. "Precatio Terrae," in *Minor Latin Poets 1*, trans. J. Wight Duff and Arnold M. Duff, Loeb Classical Library (Cambridge, MA: Harvard University Press, 1934), 342–43, ll. 1, 3.

15. Pliny, *Natural History*, trans. H. Rackham, Loeb Classical Library (Cambridge, MA: Harvard University Press, 1952), vol. IX, 33.1.

16. Pliny, *Natural History*, 33.1.

17. Carolyn Merchant, "Mining the Earth's Womb," in *Philosophy of Technology: The Technological Condition: An Anthology*, ed. Robert C. Scharff and Val Dusek (Malden, MA: Blackwell, 2003), 421–22.

18. Niavis, *Judicium Jovis*, 62.

19. Niavis, *Judicium Jovis*, 62.

20. Niavis, *Judicium Jovis*, 60.

21. Niavis, *Judicium Jovis*, 62.

22. Niavis, *Judicium Jovis*, 79–80.

23. Michel Serres, *The Natural Contract*, trans. Elizabeth MacArthur and William Paulson (Ann Arbor: University of Michigan Press, 1995), 121, 120. One might liken Serres's thought to Bill McKibben's more recent suggestion that we conceptualize a shift from Earth to Eearth in his *Eearth: Making Life on a Tough New Planet* (New York: St. Martin's Griffin, 2011).

24. Niavis, *Judicium Jovis*, 51.

25. I thank Anna Wainwright for pointing this out to me.

26. Katherine Park, "Nature in Person: Medieval and Renaissance Allegories and Emblems," in *The Moral Authority of Nature*, ed. Lorraine Daston and Fernando Vidal (Chicago: University of Chicago Press, 2010): 53, 54. On the medieval figure of Natura, see George Economou, *The Goddess Natura in Medieval Literature* (Cambridge, MA: Harvard University Press, 1972).

27. Alain of Lille, *The Complaint of Nature*, trans. Douglas M. Moffat, *Yale Studies in English* 36 (New York: Henry Holt, 1908), 15. Latin quotes are from http://www.thelatinlibrary.com/alanus/alanus1.html.

28. This connection between Alain of Lille's Natura and the Hardwick Hall portrait is made by Todd A. Borlik, who explains that "Bess of Hardwick, the Countess of Shrewsbury, commissioned the painting to commemorate her presenting this dress (which she also commissioned) as New Year's gift to the Queen. . . . Depicting Elizabeth the aging virgin as a topless mother with suckling children in her arms [one typical image of Natura] would not only have been in poor taste, it very well could have landed an artist in the Tower. But Dame Nature . . . represented a powerful female authority figure which, in a nation with a female ruler . . . provided a convenient image for naturalizing the monarch's authority over the realm." Todd A. Borlik, *Ecocriticism and Early Modern English Literature* (New York: Routledge, 2011), 69.

29. Alain of Lille, *The Complaint of Nature*, 6; Latin text, 437C, 437D.

30. Alain of Lille, *The Complaint of Nature*, 16; 452B.

31. Alain of Lille, *The Complaint of Nature*, 33.

32. Alain of Lille, *The Complaint of Nature*, 26, 465C.

33. Park, "Nature in Person," 51, 52.

34. Niavis, *Judicium Jovis*, 64–65.

35. Niavis, *Judicium Jovis*, 72.

36. Niavis, *Judicium Jovis*, 73.

37. Dipesh Chakrabarty, "The Climate of History: Four Theses," *Critical Inquiry* 35 (Winter 2009): 197–222; Bruno Latour, *Face à Gaïa. Huit conférences sur le nouveau régime climatique* (Paris: La Découverte, 2016), 154.

38. Ovid, *Metamorphoses*, trans. Frank Justus Miller, rev. G. P. Goold, Loeb Classical Library (Cambridge, MA: Harvard University Press, 1916), 1:89–106.

39. Ovid, *Metamorphoses*, 1:123–24.

40. Ovid, *Metamorphoses*, 1:125–27.

41. Ovid, *Metamorphoses*, 1:125–150.

42. Niavis, *Judicium Jovis*, 61.

43. Niavis, *Judicium Jovis*, 66. To further confirm his point, the *penates* add, later in the same *oratio*: "Locus est obscurus, immanibus saxis rupibusque" ("it is a dark place, full of huge stones and rocks") (68).

44. Niavis, *Judicium Jovis*, 74.

45. Niavis, *Judicium Jovis*, 76. This point leads to mention of *aurum potabile* (soluble gold), a gold-infused solution whose curative properties were celebrated in the Middle Ages and early modern period by authors such as Francisci Antonii (in his *Panacea aurea, sive tractatus duo de ipsius auro potabili*) and Nicholas Culpepper (in his *Treatise of Aurum Potabile*).

46. The authoritative work on this tradition is Marylène Possamaï-Pérez, *L'Ovide moralisé: essai d'interprétation* (Paris: Champion, 2006); and, by the same, this follow-up volume: *Nouvelles études sur l'Ovide moralisé* (Paris: Champion, 2009). For a short intro-

duction, see her "L'Ovide moralisé, ou la 'bonne glose' des *Métamorphoses* d'Ovide," *Cahiers d'Études Hispaniques Médiévales* 31, no. 1 (2008): 181–206.

47. See, for example, *Ovidii Quindecim Metamorphoseos libri diligentius recogniti, cum familiaribus commentariis et indice alphabetico ab Ascensio summa cura collecto* (Lyon: Jacques Huguetan, 1501), f. V^v [Gallica].

48. Ovid, *Metamorphoseos vulgare* (Venice: Z. Rosso, 1497), f. III^r: "Nel t[em]po delo inuerno se refuga et c[on]suma ogni rio humore: cosi in la terra come anche in ogni altra cosa."

49. Ovid, *Le Grand Olympe des Histoires poëtiques du prince de poësie Ouide Naso en sa Metamorphose* (Lyon: Romain Morin, 1532), f. viii^v–ix^r [Gallica].

50. See the woodcut showing the Iron Age in Ovid, *Le Grand Olympe des Histoires poëtiques*, f. viii^v.

51. On allegory and its dangers, see especially Antoine Compagnon, *Chat en poche: Montaigne et l'allégorie* (Paris: Seuil, 1993).

52. Niavis, *Judicium Jovis*, 85.

53. Niavis, *Judicium Jovis*, 87.

54. John Aberth, *An Environmental History of the Middle Ages: The Crucible of Nature* (New York: Routledge, 2012), 7–8.

55. Richard C. Hoffmann, "Homo et Natura, Homo in Natura: Ecological Perspectives on the European Middle Ages," in *Engaging with Nature: Essays on the Natural World in Medieval and Early Modern Europe*, ed. Barbara A. Hanawalt and Lisa J. Kiser (Notre Dame, IN: University of Notre Dame Press, 2008), 13, 11.

56. Florence Buttay-Jutier, *Fortuna. Usages politiques d'une allégorie morale à la Renaissance* (Paris: PUPS, 2008), 20–21. This work is by far the most nuanced and theoretically solid study of Fortune that I have come across—while revisiting the history of the changing and always multiple contents and usages of Fortune, Buttay-Jutier always recognizes and unpacks zones of uncertainty and moments of coexisting, even opposing, understandings.

57. "Fortune," in *The Classical Tradition*, ed. Anthony Grafton, Glenn W. Most, and Salvatore Settis (Cambridge, MA: Belknap Press of Harvard University Press, 2010), 365.

58. The best study of the *Consolation* remains Pierre Courcelle, *La Consolation de Philosophie dans la tradition littéraire* (Paris: Etudes Augustiniennes, 1967). On medieval and early modern reception, see Jerold C. Frakes, *The Fate of Fortune in the Early Middle Ages: The Boethian Tradition* (Leiden: Brill, 1988), esp. chapter 3 ("Boethius's Interpretation of *fortuna*"), 30–63; and Robert Black and Gabriella Pomaro, *La Consolazione della filosofia nel medioevo e nel Rinascimento italiano* (Florence: Tavarnuzze, 2000). Frakes underscores the apparent reversal that occurs in the *Consolation*: "Initially, *fortuna* is presented as the principle of disorder in the cosmos which opposes the divine *ordo*. . . . The basic opposition of order to disorder provides the essential tension in the *Consolation*. . . . Without some means of rendering this power of disorder fictive, harmless or beneficent, his system collapses. His solution includes the traditional de-deification of Fortuna and entails the astounding subjection of *fortuna* to the same system of divine order to which it has traditionally been diametrically opposed." Frakes, *The Fate of Fortune*, 63.

59. Aby Warburg, "Francesco Sassettis letzwillige Verfügung" (1907), reprinted in his *Die Erneuerung der heidnischen Antike* (Leipzig: Teubner, 1932), 1:127–58, translated into English by David Britt as "Francesco Sassetti's Last Injunction to His Sones," in Aby Warburg, *The Renewal of Pagan Antiquity* (Los Angeles: Getty Research Institute for the History of Art and the Humanities, 1999), 223–64. For a nuanced reading and critique of Warburg's position, see Buttay-Jutier, *Fortuna*, 87–161.

60. Buttay-Jutier suggests that the Renaissance image of Fortune only became widespread around 1490 (*Fortuna*, 125) and further that such images spread outside of Italy only in the first decades of the sixteenth century, e.g., to France around 1507–1509 (140).

61. A modern English translation is available as Petrarch, *Remedies for Fortune Fair and Foul*, ed. and trans. Conrad H. Rawski (Bloomington: Indiana University Press, 1991). For a useful introduction to and discussion of the reception of this text, see Buttay-Jutier, *Fortuna*, 488–90.

62. Francesco Petrarch, *De remediis utriusque fortunæ libri II* (Venice: Bernardinus Stagninus, 1536), f. 131^v–132^r and *Phisicke against fortune, aswell prosperous, as aduerse: conteyned in two bookes. Whereby men are instructed, with lyke indifferencie to remedie theyr affections, aswell in tyme of the bryght shynyng sunne of prosperitie, as also of the foule lowryng stormes of aduersitie. Expedient for all men, but most necessary for such as be subiect to any notable insult of eyther extremitie*, trans. Thomas Twyne (London: printed by Thomas Dawson for Richard Watkyns, 1579), f. 77^v–78^r.

63. The italics are those of the 1579 translation, there to make clear the text's status as quotation.

64. See Petrarch, *Von der Artzney bayder Glück des guten und widerwertigen, unnd wesz sich ain yeder inn Gelück und Unglück halten sol, auss dem Lateinischen in das Teütsch gezogen, mit künstlichen Fyguren durchausz, gantz lustig und schön gezeyret* (Gedruckt zu Augspurg durch Heynrich Steyner M. D. XXXII., 1532), f. 54^r [Gallica].

65. Aberth, *An Environmental History of the Middle Ages*, 7–8.

66. I am influenced here by Timothy Morton's reflections in "Queer Ecology," *PMLA* 125, no. 2 (2010): 273–82, that ecological thought demands not that we try to figure out if/how "we" are part of "something bigger" but that we "[work] with intimacy" in ways that avoid "organicism" (278). This is why Morton brackets off Carolyn Merchant's ecofeminist reading of Terra for its "biological essentialism" (274), which Morton sees as precluding the kinds of queer intimacies required of ecological thought, that is, because of the *cloisonnement* or shutting-outside of the Earth.

67. Félix Guattari, *The Three Ecologies*, trans. Ian Pindar and Paul Sutton (London: Continuum, 2008), 11; Latour, *Face à Gaïa*, 180.

68. Ursula K. Heise, *Sense of Place and Sense of Planet: The Environmental Imagination of the Global* (New York: Oxford University Press, 2008).

69. Bruce Clarke, "Neocybernetics of Gaia: The Emergence of Second-Order Gaia Theory," in *Gaia in Turmoil: Climate Change, Biodepletion, and Earth Ethics in an Age of Crisis*, ed. Eileen Crist and H. Bruce Rinker (Cambridge, MA: MIT Press, 2009), 294.

70. Heise, *Sense of Place and Sense of Planet*, 56.

2. TERRE'S BRILLIANT MINES

1. All quotes from the "Hymne de l'or" are from Pierre de Ronsard, *Œuvres complètes*, ed. Paul Laumonier (Paris: Droz, 1935), 8:179–205. Numbers refer to line numbers. For a good general introduction to Ronsard's *Hymnes*, see Jean Céard, *La nature et les prodiges* (Geneva: Droz, 1996), chap. 8, "Ronsard à l'écoute des signes"; the studies gathered in Madeleine Lazard, ed., *Autour des "Hymnes" de Ronsard* (Paris: Champion, 1984); as well as Philip Ford, *Ronsard's* Hymnes: *A Literary and Iconographical Study* (Tempe, AZ: Medieval and Renaissance Texts and Studies, 1997). Ronsard makes clear his project in the "Hymne de l'or" at numerous points throughout the text: "In my verses, I celebrate riches" ("je celebre en mers vers la Richesse"); "I strive / To celebrate GOLD's nobility and strength" ("je m'efforce / De celebrer de l'OR la noblesse & la force"); "I will sing the praises / Of this noble metal" ("je diray la loüenge / De ce noble metal"). Ronsard, "Hymne de l'or," v. 11, v. 16, v. 47. As Jean Frappier notes: "Ronsard a rompu selon toute apparence avec un *topos* séculaire transmis par la tradition de l'humanisme: à la malédiction lancée contre l'or il substitue un triomphe de l'or. Cette dissonance semble avoir plus ou moins déconcerté les commentateurs" ("Ronsard has by all accounts broken with a long-standing *topos* passed on by humanism. In place of the usual condemnation of gold, Ronsard substitutes the idea of gold's triumph. This dissonance seems to have more or less disconcerted his commentators"). Jean Frappier, "Tradition et actualité dans l' 'Hymne de l'or' de Pierre de Ronsard," in *Histoire, mythes et symboles: études de littérature française* (Geneva: Droz, 1976), 276.

2. Ronsard, "Hymne de l'or," v. 348. This is far from the only place that Ronsard rails against poverty and praises wealth, as explored further by Simone Perrier, "La transaction poétique chez Ronsard," in *Or, monnaie, échange dans la culture de la Renaissance. Actes du 9e Colloque international de l'Association Renaissance, humanisme, réforme*, ed. André Tournon and Gabriel-André Pérouse (Saint-Etienne: Publications de l'Université de Saint-Etienne, 1994), 199–211.

3. Jonathan Patterson, *Representing Avarice in Late Renaissance France* (Oxford: Oxford University Press, 2015), 149. Ronsard, "Hymne de l'or," v. 313–15.

4. Georgius Agricola, *Bermannus (Le mineur). Un dialogue sur les mines*, ed. Robert Halleux and Albert Yans (Paris: Les Belles Lettres, 1990). 21. English translation mine.

5. Agricola, *Bermannus*, 25.

6. Vannoccio Biringuccio, *Pirotechnia* (Venice: appresso P. Gironimo Giglio, e compagni, 1559), f. 13ʳ; Vannoccio Biringuccio, *De la Pirotechnia*, ed. and trans. Cyril Stanley Smith and Martha Teach Gnudi (New York: Dover, 1990), 26.

7. Ronsard, "Hymne de l'or," v. 173–82.

8. Ronsard, "Hymne de l'or," v. 187–95.

9. Ronsard, "Hymne de l'or," v. 277–86.

10. In Jean-Claude Margolin's words, "Ronsard ne doit ses vers qu'à son génie poétique" ("Ronsard owes these verses only to his poetic genius"). Jean-Claude Margolin, "'L'hymne de l'or' et son ambiguïté," *BHR* 28 (1966): 276. Jean Frappier ("Tradition et

actualité," 277) similarly opines, "Ronsard [ne doit cette scène mythologique] qu'à sa seule invention" ("Ronsard [owes this mythological scene] only to his own invention"), as does Terence Cave when he calls this "[un mythe] inventé de toutes pièces par Ronsard" ("a myth wholly invented by Ronsard"). Terence Cave, *Pré-Histoires II* (Geneva: Droz, 2001), 174. More recently Jonathan Parsons called this part of the poem "invented for the occasion." Jonathan Parsons, *Making Money in Sixteenth-Century France* (Ithaca, NY: Cornell University Press, 2014), 252.

11. For an introduction to Ronsard's use of mythology, see inter alia Phillip John Usher, "Introduction," in Pierre de Ronsard, *The Franciad*, ed. and trans. Phillip John Usher (New York: AMS Press, 2010), xliii–xlvii. Philip Ford (*Ronsard's* Hymnes, 159) says it stands out "in contrast to the more down-to-earth sections of the poem."

12. Emphasis mine.

13. Much has been written about Ronsard's library and reading habits. The founding article is Paul Laumonier, "Sur la bibliothèque de Ronsard," *Revue du Seizième siècle* XIV (1927): 314–35; which has been followed by many more as new discoveries have been made. The most comprehensive survey to date is in François Rouget, *Ronsard et le livre. Étude de critique génétique et d'histoire littéraire. Première Partie* (Geneva: Droz, 2010), esp. the section "Les livres annotés de Ronsard," 54–77. The work of Agricola's that Ronsard was known to have read is his *De ortu et causis* (Basel: J. Froben, 1546). Ronsard's copy is described—and a copy of the title page is included—in G. Charlier, "Un livre de la bibliothèque de Ronsard," *Revue du Seizième siècle* 9 (1921): 133–37. Unfortunately, this volume, which used to be kept at the Bibliothèque de l'Académie royale des sciences et lettres de Belgique, now seems to be missing.

14. Ulrich Rülein von Calw, *Bergbüchlein. The Little Book on Ores*, trans. Anneliese Sisco, with notes by Cyril Stanley Smith and compiled by Mary Ross (Oxshott, Surrey: Oxshott Press, 2013), 17. For an introduction to this work, see Robert Bork, ed., *De re metallica. The Uses of Metal in the Middle Ages* (Burlington: Ashgate, 2005), 347–66. For the German, see https://patrimoine.mines-paristech.fr/items/viewer/2#page/5/mode/2up.

15. Agricola, *De re metallica*, Latin 29; Eng. 43.

16. Carolyn Merchant, "Mining the Earth's Womb," in *Philosophy of Technology: The Technological Condition: An Anthology*, ed. Robert C. Scharff and Val Dusek (Malden, MA: Blackwell, 2003), 420.

17. Biringuccio, *Pirotechnia*, f. 9ᵛ; Biringuccio, *De la Pirotechnia*, 21.

18. Naomi Klein, narr., *This Changes Everything* (2015), 4:52. Merchant situates this turn with Francis Bacon, for whom Terra must be "bound into service" and "made a slave." Quoted in Merchant, "Mining the Earth's Womb," 426.

19. Jane Bennett, *Vibrant Matter: A Political Ecology of Things* (Durham, NC: Duke University Press, 2010).

20. Bennett, *Vibrant Matter*, 59. Here, Bennett builds on Gilles Deleuze, *Pure Immanence: A Life*, trans. Anne Boyman (New York: Zone, 2001).

21. Ronsard, "Hymne de l'or," v. 150. This reference is one part of what connects the "Hymne de l'or" to Ronsard's "Hymne de la Justice" in the same volume. On this aspect, see Ford, *Ronsard's Hymnes*, 153–63.

22. Ovid, *Metamorphoses*, trans. Frank Justus Miller, rev. G. P. Goold, Loeb Classical Library (Cambridge, MA: Harvard University Press, 1916), 1:102.

23. Bernardino Telesio, *De rerum natura iuxta propria principia*, ed. and trans. Arturo B. Fallico and Herman Shapiro (New York: Modern Library, 1967), 309; quoted by Merchant, "Mining the Earth's Womb," 420.

24. Ronsard, "Hymne de l'or," v. 271.

25. Ronsard, "Hymne de l'or," 77, 184–85, 106–12, 229–32.

26. I think here of metallic life as articulated by Jane Bennett, *Vibrant Matter*, 61, which builds on Deleuze's definition of "a life" as "an interstitial field of non-personal, ahuman forces, flows, tendencies, and trajectories."

27. Ronsard, "Hymne de l'or," 25–26.

28. Ronsard, "Hymne de l'or," 60.

29. Ronsard, "Hymne de l'or," 161–62.

30. Ronsard, "Hymne de l'or," 131.

31. Ronsard, "Hymne de l'or," 156.

32. Ronsard, "Hymne de l'or," 220.

33. Ronsard, "Hymne de l'or," v. 289, v. 293–94.

34. Ronsard, "Hymne de l'or," v. 295–96.

35. Ronsard, "Hymne de l'or," v. 301–02.

36. Ronsard, "Hymne de l'or," v. 48–49.

37. Pierre de Ronsard, *Le premier livre des amours*, s. 20, in *Œuvres Complètes*, ed. Paul Laumonier (Paris: Hachette, 1925), 4:23, v. 1–3. An alternative translation of these verses, which render Fr. *sein* not as "bosom" but as "lap"—Cotgrave translates "The bosome; the lap"—can be found in Pierre de Ronsard, *Selected Poems*, ed. and trans. Malcolm Quainton and Elizabeth Vinestock (London: Penguin, 2002), 4.

38. Ronsard, "Hymne de l'or," v. 168.

39. Rebecca Zorach, *Blood, Milk, Ink, Gold* (Chicago: University of Chicago Press, 2005), 195.

40. Pierre de Ronsard, *Œuvres complètes*, ed. Paul Laumonier (Paris: Hachette, 1924), 1:13, v. 74.

41. Ronsard, *Œuvres complètes*, 1:19, v. 35–36.

42. Ronsard, *Œuvres complètes*, 1:25, v. 25–26.

43. Ronsard, *Œuvres complètes*, 1:34, v. 215.

44. I base this number on Alvin Emerson Creore, *A Word-Index to the Works of Ronsard* (Leeds: W. S. Maney and Son, Ltd., 1972), 2:1338. The text in the *Hymnes* where "terre" is most frequently used is the "Hymne de la Justice," but even while the word thus necessarily refers to the place that Justice fled, it is also the place to which Justice returns. The majority of references therein are in any case rather neutral.

45. Ronsard, "Hymne de l'or," v. 389–94.

46. Ronsard, "Hymne de l'or," 196n1.

47. Stobaeus, *Sententiae ex thesauris graecorum delectae, quarum autores circiter ducentos & quinquaginta citat, & in sermones sives locos communes digestae*, trans. Conrad Gesner (Tiguri: excudebat Christophorus Froschoverus, 1543), f. 449r.

48. Agricola, *De re metallica*, 4/8.

49. Ulrich Rülein von Calw, *Bergbüchlein. The Little Book on Ores*, 36; *Ain wolgeordnetz: unnd nuezliche buchlin wie man bergwerck suche und erfinden sol von allerlay mettal die den die siben planeten generiren und wurcken* (Augsburg: Ratdolt, 1505) [École des mines de Paris, 8° Rés. 26], sig. b iv^v.

50. Rülein von Calw, *Bergbüchlein*, 36; sig. b iv^v.

51. Ronsard, "Hymne de l'or," v. 1–6.

52. Dassonville has argued ("Eléments pour une définition de l'hymne ronsardien") that each hymn is dedicated to a particularly pertinent recipient. Lancelot Carle, whose interest in occultism is well known, thus receives the "Hymne des daimons"; the author of the *Advertissement sur les jugements d'astrologie*, Mellin de Saint-Gellais, gets the "Hymne des Astres," etc. Drawing on this, Maurice Verdier noted years ago the appropriateness of the dedication of the "Hymne de l'or" to Dorat. As Verdier discusses, Dorat, like Ronsard, constantly sought sponsorship. In Verdier's words, Ronsard, who had recently enjoyed relative success at the king's table and who apostrophizes Dorat a total of six times throughout the poem, wanted to "give a lesson to [his former teacher] who feigned disdain for gold—and yet who could not live without it." Maurice Verdier, "A propos d'une controverse sur l'*Hymne de l'or* de Ronsard," *BHR* 35, no. 1 (1973): 7–18. In Verdier's reading, Ronsard's point is to quote—while turning away from—classical *topoi* on the topic of gold, in order to force attention instead on contemporary realities, namely, that gold brings power and esteem but not always to those who deserve it (12).

53. Ronsard, *Œuvres complètes*, 8:180n2. See also Victor Eugène Ardouin-Dumazet, *Voyage en France—Limousin* (Paris: Berger-Levrault & Cie., 1903), 91.

54. The contemporary in question is Papyre Masson. See Jean Dorat, *Œuvres poétiques*, ed. Charles Marty-Laveaux (Paris: A. Lemerre, 1875), vii. See also Demay, *Jean Dorat* (Paris: L'Harmattan, 1996), 23, 30.

55. Pierre de Ronsard, "Hymne de l'automne," in *Œuvres complètes*, 12:46–68, v. 78–82.

56. François de Belleforest, *La cosmographie universelle* (Paris: Michel Sonnius, 1575), col. 213.

57. Henri Deschamps, "L'or du Limousin," *La Réforme sociale. Bulletin de la société d'économie sociale et des unions de la paix sociale*, 49ème année, 9ème série, tome IX, 1re et 2e livraisons (January–February 1929): 310.

58. Belleforest, *La cosmographie universelle*, col. 209.

59. Maurice Bouguereau, *Le théâtre françoys* (Tours: Maurice Bouguereau, 1594), f. XI^r.

60. See Monique Blanc, *Emaux peints de Limoges. XVe–XVIIIe siècles. La Collection du Musée des Arts Décoratifs* (Limoges: Les Arts Décoratifs, 2011), 73.

61. John Hungerford Pollen, *Ancient and Modern Gold and Silver Smiths' Work in the South Kensington Museum* (London: Chapman and Hill, 1878), cxxix.

62. Ursula K. Heise, *Sense of Place and Sense of Planet: The Environmental Imagination of the Global* (New York: Oxford University Press, 2008).

63. Robert Robertson, "The Conceptual Promise of Glocalization," *Art-e-fac* 4, http://artefact.mi2.hr/_a04/lang_en/theory_robertson_en.htm.

64. Ronsard, "Hymne de l'or," v. 389–94.

65. Michel Peronnet, "De l'or splendeur immortelle," in *Or, monnaie, échange dans la culture de la Renaissance*, ed. André Tournon and Gabriel-André Pérouse (Saint-Etienne: Université de Sainte-Etienne, 1994), 46.

66. Ronsard, "Hymne de l'or," v. 233–42.

67. Cristina Bellorini, *The World of Plants in Renaissance Tuscany: Medicine and Botany* (Burlington, VT: Ashgate, 2016), 131.

68. Bellorini, *The World of Plants in Renaissance Tuscany*, 133. The trend would continue after Ronsard's "Hymne de l'or": The Pisan professor of anatomy Gabriele Falloppio dedicates a long section to the plant in his *Tractatus de morbo gallico* (1563), asserting that it has "an intrinsic property which cures this illness." Similarly, Andrea Cesalpino would offer a long and laudatory description of guaiacum in his *De plantis libri XVI* (1583), explaining that Columbus first came across the plant on the island of Santo Domingo (137).

69. Rebecca Earle, *The Body of the Conquistador: Food, Race, and the Colonial Experience in Spanish America, 1492–1700* (Cambridge: Cambridge University Press, 2012), 112.

3. TERRA GLOBALIZED

1. This narrative of various evolutions in early modern spatiality has become an important body of scholarship over the past several decades. Essential points of reference include the various chapters in the third volume (*Cartography in the European Renaissance*) of J. B. Harley and David Woodward, eds., *The History of Cartography* (Chicago: University of Chicago Press, 2007); as well as, inter alia, Tom Conley, *The Self-Made Map: Cartographic Writing in Early Modern France* (Minneapolis: University of Minnesota Press, 1996); Jean-Marc Besse, *Les grandeurs de la terre* (Lyon: ENS Éditions, 2003); Frank Lestringant, *L'atelier du cosmographe* (Paris: Albin Michel, 1991), available in an English translation by David Fausett as *Mapping the Renaissance World: The Geographical Imagination in the Age of Discovery* (Berkeley: University of California Press, 1994); Phillip John Usher, *Errance et cohérence: Essai sur la littérature transfrontalière* (Paris: Classiques Garnier, 2010); Phillip John Usher, *L'aède et le géographe: Poésie épique et espace du monde* (Paris: Classiques Garnier, 2018); Ayesha Ramachandran, *The Worldmakers: Global Imagining in Early Modern Europe* (Chicago: University of Chicago Press, 2015); and Katharina Piechocki, *Cartographic Humanism: Defining Early Modern Europe (1480–1580)* (forthcoming).

2. Sebastian Münster, *Cosmographiae universalis lib. VI* (Basel: apud Henrichum Petri, 1552), 11. Henceforth *Cosmographia*.

3. On the anonymity and at-a-distance-ness of the Anthropocene's *anthropos*, see Phillip John Usher, "Untranslating the Anthropocene," *Diacritics* 44, no. 3 (2016): 56–77.

4. Saba Imtiaz and Zia ur-Rehman, "Death Toll from Karachi, Pakistan, Heat Wave Tops 800," *New York Times*, June 25, 2015.

5. Pope Francis's 2015 *Laudato si'* encyclical apprehends the problem when it sets in juxtaposition the need to protect a "common home" and the fact that "exposure to atmospheric pollutants," as it "produces a broad spectrum of health hazards,"

affects "especially . . . the poor." Pope Francis, "Encyclical Letter *Laudato Si'* of the Holy Father Francis on Care for our Common Home," May 24, 2015, http://w2.vatican.va/content /francesco/en/encyclicals/documents/papa-francesco_20150524_enciclica-laudato-si .html. The encyclical makes similar points throughout, for example noting that "Many of the poor live in areas particularly affected by phenomena related to warming, and their means of subsistence are largely dependent on natural reserves and ecosystemic services such as agriculture, fishing and forestry." A vast and growing body of literature about climate justice now exists. Starting points include Eric A. Posner and David Weisbach, *Climate Change Justice* (Princeton, NJ: Princeton University Press, 2010); James B. Martin-Schramm, *Climate Justice: Ethics, Energy, and Public Policy* (Minneapolis: Fortress, 2010); Vandana Shiva, *Soil Not Oil: Environmental Justice in a Time of Climate Crisis* (Cambridge, MA: South End, 2008). For postcolonial perspectives, see Dipesh Chakrabarty, "Postcolonial Studies and the Challenge of Climate Change," *NLH* 2012 (43): 1–18; and Ian Baucom, "The Human Shore: Postcolonial Studies in an Age of Natural Science," *History of the Present* 2, no. 1 (Spring 2012): 1–23.

6. Peter H. Meurer, "Cartography in the German Lands, 1450–1650," in *The History of Cartography*, ed. David Woodward (Chicago: University of Chicago Press, 2007), vol. 3, part 2, 1213.

7. Münster, *Cosmographia*, 11.

8. Münster, *Cosmographia*, 12. On medieval cartography, see especially David Woodward, "Medieval Mappaemundi," in *History of Cartography*, vol. 1, *Cartography in Prehistoric, Ancient, and Medieval Europe and the Mediterranean*, ed. J. B. Harley and David Woodward (Chicago: University of Chicago Press, 1987), 286–370; as well as Evelyn Edson, *Mapping Time and Space: How Medieval Mapmakers Viewed Their World* (London: British Library, 1997); Evelyn Edson, *The World Map, 1300–1492: The Persistence of Tradition and Transformation* (Baltimore, MD: Johns Hopkins University Press, 2007). On some of the ways in which early modern mapping demonstrated (as here in Münster) a certain nostalgia vis-à-vis earlier medieval modes, see Phillip John Usher, "The Holy Lands in Early Modern Literature: Negotiations of Christian Geography and Textual Space," PhD diss., Harvard University, 2004.

9. Münster, *Cosmographia*, 1099. "Münster allocated 18 percent of the *Cosmographia* to Britain, Spain, France and Italy; 15 percent to the kingdoms of north; 12 percent to Asia and the New World; and 4 percent to Africa. To the German lands he gave 48 percent of his book." Matthew McLean, *The Cosmographia of Sebastian Münster: Describing the World in the Reformation* (Abingdon: Routledge, 2007), 193.

10. Münster, *Cosmographia*, 6–10.

11. Further on in the work, Münster will give over significant space—many detailed and richly illustrated pages—to extraction in his home country, Germany (*Cosmographia*, 430–37). McLean (*The Cosmographia of Sebastian Münster*, 199–200) notes the nationalistic tone of these pages: "Attached to this description are passages recalling to the reader the especial blessedness of Germany in terms of minerals and metals; the ingenuity of recent mining methods; their method for the location of deposits; the remarkable boom caused by the discovery of a rich vein following the Peasants'

War which caused 1,200 new houses to be built in the mountains. Maps illustrate the locations of the main mining sites, cutaway diagrams show the mining process, and, in passing, a small illustration shows a 'miracle in nature': a creature shaped and depicted in the mineral-rocks, which some aver to be a fish, some a reptile and others a frog." In his pages on the New World, Münster will frequently mention gold: "Gold is considered by [New World inhabitants] to be priceless" ("Aurum apud eos est in praecio") (*Cosmographia*, 1100); "This region abounds in gold, parrots, silkworms, and incredibly good weather" ("Ea regio abundat auro, psitacis, bombice, & incredibili temperie aëris") (1107).

12. All quotes are from Michel de Montaigne, *Essais*, ed. Pierre Villey and V.-L. Saulnier (Paris: PUF, 1965); and *The Complete Essays*, trans. M. A. Screech (London: Penguin, 1991). Here Montaigne, *Essais*, III, 6, 911/1031.

13. Montaigne, *Essais*, 910/1031.

14. Richard Anthony Sayce, *The Essays of Montaigne* (London: Weidenfeld and Nicolson, 1972), 273.

15. Montaigne, *Essais*, III, 6, 915/1036–37.

16. Montaigne likely reverses the order of events for "thematic reasons," that is, "so that the episode of the litter comes at the end, after he has already recounted [Atahualpa's] death," such that this final episode about a coach marks the end, just as Montaigne's discussion of his own use of coaches marks the beginning, thus providing a coherent frame. Deborah N. Losse, *Montaigne and Brief Narrative Form* (New York: Palgrave Macmillan, 2013), 81. The chapter's overall design (or at least structure) has been described by Edwin Duval in his "Lessons of the New World: Design and Meaning in Montaigne's 'Des cannibales' (I:31) and 'Des coches'' (III:6)," *Yale French Studies* 64 (1983): 95–112.

17. Timothy Hampton, "The Subject of America: History and Alterity in Montaigne's 'Des coches,'" *Romanic Review* 88 (March 1997): 221.

18. Tom Conley, "The *Essays* and the New World," in *The Cambridge Companion to Montaigne*, ed. Ullrich Langer (Cambridge: Cambridge University Press, 2005): 89.

19. François Lopez de Gómara, *Histoire generalle des Indes occidentales, et terres neuves, qui iusques à present ont esté descouuertes* (Paris: chez Michel Sonnius, 1584), f. 311ᵛ.

20. For differing points of view, see Marcel Bataillon, "Montaigne et les conquérants de l'or," *Studi Francesi* 9 (1959): 353–67; Géralde Nakam, *Les Essais de Montaigne. Miroir et procès de leur temps* (Paris: Nizet, 1984), 340; Géralde Nakam, *Montaigne et son temps* (Paris: Nizet, 1982), 40–41; Juan Durán Luzio, "Bartolomé de las Casas y Michel de Montaigne: Escritura y lectura del Nuevo Mundo," *Revista Chilena de Literatura* 37 (April 1991): 7–24.

21. Bartolomé de las Casas, *Brevísima relación de la destrucción de las Indias*, ed. José Miguel Martínez Torrejón (Antioquia: Editorial Universidad de Antioquia, 2006), 42; *A Short Account of the Destruction of the Indies*, trans. Nigel Griffin (London: Penguin, 1999), 31–32.

22. Niavis, *Judicium Jovis*, 87. Quoted and discussed in Chapter 1.

23. As Las Casas notes later, from the formerly heavily populated coastal area of the Paria Peninsula, "over two million souls" were kidnapped and shipped to Hispaniola

and Puerto Rico "to be sent down the mines or put to other work" (*Brief Account*, 92). In Cuba, one Spanish official worked "three hundred natives . . . so hard" that at the end of three months "only thirty—that is to say, just one tenth of the original number—were still alive, the other two hundred and seventy having perished down the mines" (30). And the activity produced collateral damage in miners' families: "More than seven thousand children died of hunger, after their parents had been shipped off to the mines" (30). When Las Casas concludes the chapter on Cuba by referring to the fact that the "whole island was devastated and depopulated" and how this resulted in a "moving and heartrending spectacle, transformed, as it had been, into one vast, barren wasteland," the destroyed landscape is both human and nonhuman (30). The same *would* have happened in the "Kingdom of Yucatán": "If there had been any gold in this province, the Spaniards would have finished off the local population by sending them down mines to dig it out of the earth" (72). Las Casas's text, as it follows from place to place the consequences of European presence in the New World, makes the extraction of precious metals both a goal, a means, and a metaphor of the Spanish colonial project.

24. Las Casas, *Brevísima relación*, 118; *A Short Account*, 97.

25. Las Casas, *A Short Account*, 83.

26. Las Casas, *A Short Account*, 125.

27. Las Casas, *A Short Account*, 123.

28. Las Casas, *A Short Account*, 123.

29. Las Casas, *A Short Account*, 97.

30. Las Casas, *A Short Account*, 100.

31. http://www.museoroperu.com.pe/oro.html.

32. Raquel Gil Montero, "Free and Unfree Labour," *IRSH* 56 (2011): 297.

33. Nicholas Robins, *Mercury, Mining, and Empire: The Human and the Ecological Cost of Colonial Silver Mining in the Andes* (Bloomington: Indiana University Press, 2011), 75.

34. Pedro de Cieza de León, *Parte primera de la Chronica del Perv, qve tracta la demarcacion de sus prouincias, la descripcion dellas, las fundaciones de las nueuas ciudades, los ritos y costumbres de los Indios, y otras cosas estrañas dignas de ser sabidas* (Antwerp: En casa de I. Steelsio, 1554), 260.

35. See Edmund Spenser, *The Faerie Queene*, ed. A. C. Hamilton (Harlow: Longman, 2001), 5.2.47; Richard Kagan, *Urban Images of the Hispanic World, 1493–1793* (New Haven, CT: Yale University Press, 2000), 103.

36. Pedro de Cieza de León, *The Travels [Chronica del Peru]*, trans. and ed. Clements R. Markham (New York: B. Franklin, 1964), 13.

37. Pedro de Cieza de León, *The Travels*, 386.

38. For a general introduction to these images and further bibliographical references, see Henry Keazor, "Théodore de Bry's Images for America," *Print Quarterly* 15, no. 2 (June 1998): 131–49.

39. For a first glimpse at the photography of Burtynsky, see https://www.edward burtynsky.com/. For his extraction landscapes in particular, see Edward Burtynsky, *Quarries: The Quarry Photographs of Edward Burtynsky* (Göttingen: Steidl, 2007).

40. Robins, *Mercury, Mining, and Empire*, 75.

41. Robins, *Mercury, Mining, and Empire*, 75.

42. Robins, *Mercury, Mining, and Empire*, 78. Of the ninety laws in Viceroy Toledo's *Ordenancas* of 1574, only three concerned questions of safety.

43. Robins, *Mercury, Mining, and Empire*, 78.

44. Robins, *Mercury, Mining, and Empire*, 74.

45. Jane E. Mangan, *Trading Roles: Gender, Ethnicity, and the Urban Economy in Colonial Potosi* (Durham, NC: Duke University Press, 2005), 4.

46. Montaigne, *Essais*, III, 6, 915/1036.

47. Montaigne, *Essais*, III, 6, 915/1036.

48. Ramiro Matos Mendieta and Jose Barreiro, eds., *The Great Inka Road: Engineering an Empire* (Washington, DC: Smithsonian Books, 2015), 110.

49. Matos Mendieta and Barreiro, eds., *The Great Inka Road*, 111.

50. Montaigne, *Essais*, III, 6, 911/1031.

51. These four declarations form the "Requirement." See Patricia Seed, *Ceremonies of Possession in Europe's Conquest of the New World, 1492–1640* (Cambridge: Cambridge University Press, 1995), chap. 3, "The Requirement: A Protocol for Conquest."

52. Montaigne, *Essais*, III, 6, 911/1032.

53. Whether or not Montaigne *intended* us to read for this homophonic and -graphic play is, from my perspective, not particularly pertinent. Apart from anything else, we can never *know* Montaigne's intention here. Among modern critics, one at least states without hesitation that Montaigne probably was cognizant of the play. There is a "word play on *mine*": "The Spanish are searching for 'mines' in the sense of 'ore' [and] the Native Americans respond using the same word to mean 'facial expression.'" "Montaigne thus bends the use of language . . . French readers are not only linguistically separated from the Spanish, but also firmly in the corner of the Native Americans as the latter, with Montaigne's help, verbally make use of a French play on words to mock the invaders." Thomas Parker, "Art and Nature: The Old and New World Seen through Montaigne's Spanish Mirror," *Montaigne Studies* 22 (2010): 35.

54. Quoted in Naomi Klein, *This Changes Everything* (New York: Simon & Schuster, 2014), 161.

PART II. WELCOME TO MINELAND

1. See Bruno Latour, *Où atterrir?* (Paris: La Découverte, 2017), 39–45; as well as the exhibition held at the Musée de la chasse et de la nature in Paris titled "Animer le paysage," on which one can consult issue no. 10 of *Billebaude* (2017), *Sur la piste des vivants*.

2. Latour, *Où atterrir?*, 57, 56, 95.

3. In Latourian terms, the emphasis is again away from "Galilean objects" and toward "Lovelockian agents." Latour, *Où atterrir?*, 87–88.

4. Philippe Quesne, *Welcome to Caveland!* (Nanterre: Nanterre-Amandiers, 2016), 3. In French: "des sous-sols et des cavernes, des territoires *underground* et alternatifs, proches de la terre et de la matière, mais aussi propices aux visions oniriques."

5. Phillip John Usher, "Welcome to Caveland! From the Exterranean to the Inter-ranean, and Then Some," November 6, 2016, http://thehumanistanthropocene.weebly.com/blog/welcome-to-caveland-from-the-exterranean-to-the-interranean-and-then-some.

6. I should like to thank Frédérique Aït-Touati, Emmanuele Coccia, and Alice Leroy for inviting me to speak in their seminar on "La connaissance sensible" at the EHESS on March 27, 2017, on the topic of "Les mondes souterrains," alongside Philippe Quesne, who spoke about *Welcome to Caveland!* I also thank Philippe Quesne for a productive and enjoyable dialogue.

7. Bruno Latour, *We Have Never Been Modern*, trans. Catherine Porter (Cambridge, MA: Harvard University Press, 1993), chap. 2.

4. SICKLY MOUNTAINSIDES

1. Throughout, I will be quoting from this edition: Georgius Agricola, *De re metallica* (Basel: Froben, 1556). Unless otherwise noted, English translations are from Georgius Agricola, *De re metallica*, trans. Herbert Clark Hoover and Lou Henry Hoover (New York: Dover, 1950).

2. Gray A. Brechin, *Imperial San Francisco: Urban Power, Earthly Ruin* (Berkeley: University of California Press, 1999), 25.

3. Eric H. Ash, *Power, Knowledge, and Expertise in Elizabethan England* (Baltimore, MD: Johns Hopkins University Press, 2004), 23. Ash writes that Agricola's book was the "single most important work on mining operations produced during the early modern period."

4. The *De re metallica*, say the editors of the recent Italian translation thereof, "si inserisce nella tradizione dei trattati di carattere tecnico" ("belongs to the tradition of technical treaties"). Paolo Macini and Ezio Mesini, "Introduzione," in Georgius Agricola, *De re metallica (1530–1556). Un dialogo sul mondo minerale e un trattato sull'arte de' metalli*, ed. Paolo Macini and Ezio Mesini (Bologna: Clueb, 2008), 14. They also note that the work was widely read and translated precisely because of "il grande interesse tecnico e scientifico che suscitò sin dai primi anni dalla sua apparizione" ("the great technical and scientific interest to which it gave rise upon publication") (15).

5. The split between the reception of Boyle and Hobbes is at the heart of Bruno Latour, *We Have Never Been Modern*, trans. Catherine Porter (Cambridge, MA: Harvard University Press, 1993).

6. Bruno Latour, *Où atterrir?* (Paris: La Découverte, 2017), 59. French: "Tous les traits d'un 'cadre' à l'intérieur duquel on pouvait . . . distinguer sans trop de peine l'action des humains, de même qu'au théâtre on peut oublier le bâtimnet et les coulisses pour se concentrer sur l'intrigue."

7. The standard account of Agricola's life is provided by Helmut Wilsdorf, *Georg Agricola und seine Zeit* (Berlin: Deutscher Verlag der Wissenschaften, 1956). See also Owen Hannaway, "Georgius Agricola as Humanist," *Journal of the History of Ideas* 53, no. 4 (October–December 1992): 553–60.

8. The following details about the growth of St. Joachimstahl and the changes in the Erzgebirge are drawn from Jiri Majer, "Ore Mining and the Town of St. Joachimsthal/Jáchymov at the time of Georgius Agricola," *GeoJournal* 32, no. 2 (February 1994): 91–99.

9. *De ortu et causis subterraneorum* (1546) discusses physical geology (wind, water, earthquakes, volcanic eruptions); *De natura fossilium* (1546) is an important contribution to paleontology—here *fossil* has a wider sense than today, referring to anything dug out of the ground; *De veteribus et novis mettalis* (1546) deals with the correspondence between ancient and modern mining sites.

10. See the recent French edition of the *Bermannus* prepared by Robert Halleux and Albert Yans (Paris: Les Belles Lettres, 1990) as well as their useful introduction.

11. On Ramism, see Anthony Grafton and Lisa Jardine, *From Humanism to the Humanities* (Cambridge, MA: Harvard University Press, 1986), especially chap. 7, "Pragmatic Humanism: Ramism and the Rise of the Humanities."

12. Macini and Mesini, "Introduzione," 15. In Italian: "difesa della professione."

13. Bern Dibner, *Agricola on Metals* (Norwalk, CT: Burndy Library, 1958), 29.

14. Agricola, *De re metallica*, 8/12.

15. Agricola, *De re metallica*, 5/8.

16. "Kettle logic" (*la logique du chaudron*) is the term Jacques Derrida uses in reference to Freud's "kettle story," as told in the *Interpretation of Dreams* and *Jokes and Their Relation to the Unconscious*. See Jacques Derrida, *Resistances of Psychoanalysis*, trans. Peggy Kamuf, Pascale-Anne Brault, and Michael Naas (Stanford, CA: Stanford University Press, 1998), 6–7.

17. Agricola, *De re metallica*, 10/14.

18. Agricola, *De re metallica*, 5/9.

19. Agricola, *De re metallica*, 10/14.

20. Agricola, *De re metallica*, f. a3^r/xxix.

21. A work rediscovered by Poggio Bracciolini in 1411 and printed for the first time in 1486.

22. Vitruvius, *The Ten Books of Architecture*, trans. Morris Hicky Morgan (New York: Dover, 1960), 1.1.1; Agricola, *De re metallica*, 1/3. Emphasis mine.

23. Agricola, *De re metallica*, 4.

24. Agricola, *De re metallica*, 4.

25. Agricola, *De re metallica*, 1/4.

26. Agricola, *De re metallica*, 1/4.

27. Agricola, *De re metallica*, 1/4.

28. Agricola, *De re metallica*, 1/4.

29. Agricola, *De re metallica*, 1/3.

30. Stobaeus, *Sententiae ex thesauris graecorum delectae, quarum autores circiter ducentos & quinquaginta citat, & in sermones sives locos communes digestae*, trans. Conrad Gesner (Tiguri: excudebat Christophorus Froschoverus, 1543), f. 449r. Agricola, *De re metallica*, 4/8. See Chapter 2 in the present volume.

31. Agricola, *De re metallica*, 4/6.

32. Agricola, *De re metallica*, 4/7. I quote the English translation from Ovid, *Metamorphoses*, trans. Frank Justus Miller, rev. G. P. Goold (Cambridge, MA: Harvard University Press, 1916), 1:137–43.

33. Agricola, *De re metallica*, 8/12.

34. Agricola, *De re metallica*, 8/12.

35. Agricola, *De re metallica*, 8/12.

36. Agricola, *De re metallica*, 8/12.

37. Agricola, *De re metallica*, 8/12.

38. Mitchell Thomashow, *Bringing the Biosphere Home: Learning to Perceive Global Environmental Change* (Cambridge, MA: MIT Press, 2002), 7.

39. Latour, *Où atterrir?*, 74. French: "Le Terrestre tient à la terre et au sol mais il est aussi mondial, en ce sens qu'il ne cadre avec aucune frontière, qu'il déborde toutes les identités."

40. On the production of these woodcuts, see in particular Pamela O. Long, "Of Mining, Smelting, and Printing: Agricola's *De re metallica*," *Technology and Culture* 44, no. 1 (January 2003): 100–1.

41. Agricola, *De re metallica*, title page. Emphasis mine.

42. Agricola, *De re metallica*, sig. a3ᵛ/xxx. Agricola further compares the potential problem for posterity to the difficulty that humanists experienced because of the "many names that the Ancients . . . have handed down to us without any explanation."

43. Pamela H. Smith, "Art, Science, and Visual Culture in Early Modern Europe," *Isis* 97, no. 1 (March 2006): 83–100.

44. Owen Hannaway, "Reading the Pictures: The Context of Georgius Agricola's Woodcuts," *Nuncius* 12, no. 1 (1997): 54.

45. Hannaway, "Reading the Pictures," 57.

46. Marie-Claude Déprez-Masson, "Les techniques de l'image," in *Technique, mot et image: Le De re metallica d'Agricola* (Turnhout: Brepols, 2006), 171–72. Translation mine.

47. Déprez-Masson, "Les techniques de l'image," 171–72.

48. Agricola, *De re metallica*, 21/33.

49. Agricola, *De re metallica*, 22/32.

50. Agricola, *De re metallica*, 22/32.

51. Agricola, *De re metallica*, 26/38.

52. Agricola, *De re metallica*, 26/38.

53. Agricola, *De re metallica*, 26/38.

54. Agricola, *De re metallica*, 26/38.

55. Michel Serres, *The Natural Contract*, trans. Elizabeth MacArthur and William Paulson (Ann Arbor: University of Michigan Press, 1995), 1.

56. Serres, *The Natural Contract*, 2.

57. Serres, *The Natural Contract*, 3.

58. Agricola, *De re metallica*, 27–28/39–40.

59. Agricola, *De re metallica*, 28/39.

60. Agricola, *De re metallica*, 28/40–41. The reference to Ulysses is to Homer, *Odyssey*, trans. A. T. Murray, rev. George E. Dimock, Loeb Classical Library (Cambridge, MA: Harvard University Press, 1995), 16:172.

61. On Münster's illustration of a divining rod in the 1550 edition of the *Cosmographia*, see Christopher Bird, *The Divining Hand: The Art of Searching for Water, Oil, Minerals, and Other Natural Resources or Anything Lost, Missing, or Badly Needed* (New York: Dutton, 1979), 16.

62. Agricola, *De re metallica*, 26/38.

63. Agricola, *De re metallica*, 28/41.

64. Jane Bennett, *Vibrant Matter: A Political Ecology of Things* (Durham, NC: Duke University Press, 2010), 14.

65. See Levi Byrant, *Onto-Cartography: An Ontology of Machines and Media* (Edinburgh: Edinburgh University Press, 2014).

66. I borrow the reference to ecomimesis from Timothy Morton, *Ecology without Nature: Rethinking Environmental Aesthetics* (Cambridge, MA: Harvard University Press, 2007).

67. William Sherman, *Used Books: Marking Readers in Renaissance England* (Philadelphia: University of Pennsylvania Press, 2008).

68. See Sherman, *Used Books*, 51: "So the hand is not just the 'instrument of instruments' (as Aristotle called it) but also the sign of signs, when it comes to learning how to communicate and representing our relationship to the tools we use. For Heidegger, indeed, this handiness (and not just our opposable thumbs) plays a key role in defining what makes us human."

69. Quoted in Sherman, *Used Books*, 51.

5. DEMONIC MINES

1. See Ignacio Miguel Pascual Valderrama and Joaquín Pérez-Pariente, "Alchemy at the Service of Mining Technology in Seventeenth-Century Europe, According to the Works of Martine de Bertereau and Jean du Chastelet," *Bulletin for the History of Chemistry* 37, no. 1 (2012): 5. The most recent edition of Kircher's text is Athanasius Kircher, *Mundus subterraneus* (Wien: Im Selbstverlag, 2005).

2. To the best of my knowledge, no such complete cultural history exists in a single volume. As a starting point, see Ronald M. James, "Knockers, Knackers, and Ghosts: Immigrant Folklore in the Western Mines," *Western Folklore* 51, no. 2 (April 1992): 153–77.

3. Jane Bennett, *Vibrant Matter: A Political Ecology of Things* (Durham, NC: Duke University Press, 2010), 31.

4. As we shall see in due course in more detail, the *De animantibus subterraneis* would also be included in the 1556 edition of the *De re metallica*. For the Latin text, I shall thus refer to and quote from the same edition of the *De re metallica* used up until this point. For the English translation, I shall quote Lindsay L. Sears in Michele L. Aldrich, Alan E. Leviton, and Lindsay L. Sears, "Georgius Agricola, *De animantibus subterraneis*, 1549 and 1566: A Translation of a Renaissance Essay in Zoology and Natural History," *Proceedings of the California Academy of Sciences* 60, no. 9 (May 7, 2009): 89–174, trans. 97–122. In writing, I have also made use of the French translation provided as "Le livre des créatures qui vivent sous terre," in Georgius Agricola, *De re metallica*, trans. Albert France-Lanord (Thionville: Gérard Klopp, 1992), 481–508.

5. Marie-Claude Déprez-Masson, *Technique, mot et image: Le* De re metallica *d'Agricola* (Turnhout: Brepols, 2006), 65.

6. Agricola, *De animantibus subterraneis*, 479/98.

7. On *De natura eorum quae effluent ex terra*, see Déprez-Masson, *Technique, mot et image*, 51.

8. Georgius Agricola, *De natura fossilium* (*Textbook of Mineralogy*), ed. and trans. Mark Choice Bandy and Jean A. Bandy (New York: Geological Society of America, 1955).

9. Agricola, *De animantibus subterraneis*, 499/119.

10. Moles, indeed, are the ultimate subterranean animal. Agricola describes them as follows: "This quadruped is not much different from a mouse, except it is blind, although it has the likeness of eyes, as Pliny writes, if anyone should remove the membrane stretched over them. It is not covered with hair. It can even hear when it is submerged in water, but when removed from the earth, which it inhabits here and there in fields and even more often in meadows and gardens, it can't live for very long. It has short legs, because of which it walks slowly. There are five toes on each of its front feet, four on the back feet, all are armed with sharp claws, with which it digs in the earth. Moreover, it has hair marked with glittering black patches [that] are white on the young of this species. It eats frogs, even the poisonous ones, earthworms, and the roots of crops and grasses. The pelts of these animals become felt caps and bed-covers." *De animantibus subterraneis*, 499–500/119.

11. Agricola, *De animantibus subterraneis*, 501–2/121. Translation modified.

12. Agricola, *De animantibus subterraneis*, 502/121. Translation modified. Emphasis mine.

13. Emanuele Coccia, *La vie des plantes. Une métaphysique du mélange* (Paris: Rivages, 2016), 23. Translation mine.

14. See, in particular, Bruno Latour, *Reassembling the Social: An Introduction to Actor-Network-Theory* (Oxford: Oxford University Press, 2005), 1–17. I thank one of my anonymous readers for pointing me in this useful direction.

15. Agricola, *De animantibus subterraneis*, 502/121. Translation modified.

16. Agricola, "Le livre des créatures qui vivent sous terre," 507.

17. This is clearly how the text's English translator understands the situation. What I render as "although sometimes the demons harass the miners with extracted matter, only rarely however do they hurt them," the published English version, which takes liberties with the Latin, reads as follows: "Although in fact the dirt sometimes irritates the workers . . ." Agricola, *De animantibus subterraneis*, 502/121. An extract from this part of the book is provided in translation in a footnote to a passage in the *De re metallica*, which I quote here for comparison: "Sometimes [the demons] throw pebbles at the workmen, but they rarely injure them unless the workmen first ridicule or curse them." Agricola, *De re metallica*, 217n26.

18. Agricola, *De animantibus subterraneis*, 502/122.

19. Agricola, *De re metallica*, 170–72/212–14.

20. Agricola, *De re metallica*, 172/214.

21. Agricola, *De re metallica*, 172/214.

22. Agricola, *De re metallica*, 173/217.

23. Agricola, *De re metallica*, 173/217.

24. Agricola, *De re metallica*, 173/217.

25. Agricola, *De re metallica*, 173–74/217–18.

26. Henry Heller, *Labour, Science, and Technology in France, 1500–1620* (Cambridge: Cambridge University Press, 1995), 130–31.

27. See François Garrault, *Paradoxe sur le faict des monnoyes* (Paris: J. Du Puy, 1578), sig. Av^v.

28. François Garrault, *Recueil des principaux advis* (Paris: Chez Jacques du Puys, 1578), 8.

29. Garrault, *Recueil des principaux advis*, 14.

30. François Garrault, *Des mines d'argent trouvées en France, ouvrage & police d'icelles* (Paris: pour la Veuve Jehan Dalier & Nicolas Rosset, 1579), f. Aii^r.

31. Garrault, *Des mines d'argent trouvées en France*, f. Aii^r–Aii^v.

32. Garrault, *Des mines d'argent trouvées en France*, f. Aii^v.

33. Garrault, *Des mines d'argent trouvées en France*, f. Aii^v.

34. Amédée Burat, *Géologie appliquée, ou traité de la recherche de l'exploitation des minéraux utiles* (Paris: Langlois et Leclercq, 1858–1859), 177.

35. Burat, *Géologie appliquée*, 149.

36. James Dwight Dana, *Manual of Mineralogy* (New York: Wiley, 1985), 274–76.

37. Garrault, *Des mines d'argent trouvées en France*, f. C. iii^r.

38. Garrault, *Des mines d'argent trouvées en France*, f. Ciii^v.

39. Randle Cotgrave, *A Dictionarie of the French and English Tongues* (London: Adam Islip, 1611).

40. Garrault, *Des mines d'argent trouvées en France*, f. Ciii^v.

41. Garrault, *Des mines d'argent trouvées en France*, f. Ciii^v, f. Civ^r.

42. Garrault, *Des mines d'argent trouvées en France*, f. Civ^r.

43. Garrault, *Des mines d'argent trouvées en France*, f. Eiii^r.

44. Garrault, *Des mines d'argent trouvées en France*, f. Eiv^r.

45. Paracelsus, *Von der Bergsucht* (Dilingen: Mayer, 1567), 3^r; Paracelsus, *On the Miners' Disease*, bk. 1, tract. 1, chap. 2, in *Four Treatises*, ed. Henry E. Sigerist (Baltimore, MD: Johns Hopkins University Press, 1996), 58. A general introduction to Paracelsian medicine is provided by Walter Pagel, *Paracelsus: An Introduction to Philosophical Medicine in the Era of the Renaissance* (New York: Krager, 1982), esp. 126–52; and Bruce Moran, *Distilling Knowledge: Alchemy Chemistry, and the Scientific Revolution* (Cambridge, MA: Harvard University Press, 2005), esp. chap. 3, "Paracelsus and the 'Paracelsians': Natural Relationships and Separation as Creation."

46. Paracelsus, *On the Miners' Disease*, 1.1.3, 59.

47. Paracelsus, *On the Miners' Disease*, 1.1.3, 60.

48. Paracelsus, *On the Miners' Disease*, 1.1.3, 60.

49. Paracelsus, *On the Miners' Disease*, 1.1.4, 60–61.

50. Paracelsus, *On the Miners' Disease*, 1.1.4, 61–62; 1.2.3, 63; 1.3.3, 71.

51. Paracelsus, *Von der Bergsucht*, 11^v; Paracelsus, *On the Miners' Disease*, 1.3.1, 67.

52. Paracelsus, *On the Miners' Disease*, 1.3.2, 57.

53. Paracelsus, *On the Miners' Disease*, 1.1.1, 73.

54. Paracelsus, *On the Miners' Disease*, 1.4.2, 74.

55. Bennett, *Vibrant Matter*, ix.

56. Bennett, *Vibrant Matter*, 18.

PART III. HIDING IN EXTERRANEAN MATTER

1. The term "Earth-bound" is used inter alia by Bruno Latour, e.g., in "An Attempt at a 'Compositionist Manifesto,'" *New Literary History* 41 (2010): 472.

2. The idea of intra-action belongs to Karen Barad, *Meeting the Universe Halfway* (Durham, NC: Duke University Press, 2007).

3. The term *sympathy* derives from ancient Greek σύν (together) and πάθος (feeling). For a survey of the term's usages in classical and early modern contexts, see Eric Schliesser, ed., *Sympathy: A History* (Oxford: Oxford University Press, 2015). The rallying cry to "make kin" is from Donna Haraway, "Anthropocene, Capitalocene, Plantationocene, Chthulucene: Making Kin," *Environmental Humanities* 6 (2015): 159–65.

4. As Philippe Dagen put it in relation to Bolin's series as a whole, "forms of pollution" and the "destruction of ecosystems" are shown to be occurring in plain sight. Philippe Dagen, "Introduction," in *Liu Bolin*, ed. Laura Dozier (New York: Abrams, 2014), 8. That Bolin's series tackles ecology head-on is clearly also Bolin's intention: "We control the world and it is we who decide if we treat our environment friendly [sic] or not. If we treat the environment as a friend, it will improve our lives. If we treat the environment as an enemy, we will destroy it. We must resolve this dilemma." Liu Bolin, *The Invisible Man* (Hong Kong: Blue Kingfisher, 2011), n.p.

5. Jeffrey Jerome Cohen, *Stone: An Ecology of the Inhuman* (Minneapolis: University of Minnesota Press, 2015); Tiffany Werth, "Loving London Stone," *Upstart: A Journal of Renaissance English Studies*, February 14, 2014, http://www.clemson.edu/upstart/Essays/london_stone/london_stone.xhtml.

6. GEOMEDIA

1. The Anthropocene's invitation to reconnect human history and geological time is elaborated and studied by Dipesh Chakrabarty in his "The Climate of History: Four Theses," *Critical Inquiry* 35 (Winter 2009): 197–222.

2. The work of Jussi Parikka here is paramount, especially his *What Is Media Archaeology?* (Malden, MA: Polity, 2012) and *A Geology of Media* (Minneapolis: University of Minnesota Press, 2015). On Bruce Sterling's Dead Media Project, see http://www.deadmedia.org. I think also of Erik Hagen's exhibition *Fossils of the Anthropocene* at the AAAS Art Gallery in 2014.

3. Ian Sample, "Google Boss Warns of 'Forgotten Century,'" *Guardian*, February 13, 2015. The article quotes Cerf as saying something that, from the vice president of

Google, should make us wake up and pay attention: "If there are photos you really care about, print them out."

4. Tiffany Werth, "Loving London Stone," *Upstart: A Journal of Renaissance English Studies*, February 14, 2014, http://www.clemson.edu/upstart/Essays/london_stone/london_stone.xhtml.

5. Jeffrey Jerome Cohen, *Stone: An Ecology of the Inhuman* (Minneapolis: University of Minnesota Press, 2015), 21.

6. On "things" in the phenomenon of hoarding, see Jane Bennett's lecture "Powers of the Hoard: Artistry and Agency in a World of Vibrant Matter," https://www.youtube.com/watch?v=q607Ni23QjA.

7. See Olivier Dugué, Laurent Dujardin, Pascal Lerous, and Zavier Savary, *La pierre de Caen: Des dinosaures aux cathédrales* (Condé-sur-Noireau: Editions Charles Corlet, 2010), chap. 1, "Géologie du Calcaire de Caen," 11–36.

8. Dugué et al., *La pierre de Caen*, 24. The village where the stone was mined is now called Fleury-sur-Orne.

9. At this time, France was what geologists call a platform, a more or less flat surface covered by seawater and onto which successive layers of sediment are laid down. At this point, only three "islands" emerge. For more details, see Dugué et al., *La pierre de Caen*, 12.

10. On the idea of the Earth as a recording machine, see John Durham Peters, "Space, Time, and Communication Theory," *Canadian Journal of Communication* 28, no. 4 (2003), http://www.cjc-online.ca/index.php/journal/article/view/1389/1467.

11. John Ashurst and Francis G. Dimes, *Conservation of Building and Decorative Stone* (Abingdon: Routledge, 2011), 2:121.

12. On early modern architecture in France, see Frédérique Lemerle and Yves Pauwels, *L'architecture à la Renaissance* (Paris: Flammarion, 1998); Anthony Blunt, *Art and Architecture in France, 1500–1700* (New York: Penguin, 1981); and Henri Zerner, *L'art de la Renaissance en France: L'invention du classicisme* (Paris: Flammarion, 1996).

13. For an introduction to the early modern Louvre, see Zerner, *L'art de la Renaissance en France*, 157–60; as well as Christiane Aulanier, "Le Palais du Louvre au XVIᵉ siècle: Documents inédits," *Bulletin de la Société de l'histoire de l'art français* (1951): 85–100; and Phillip John Usher, "From Marriage to Massacre: The Louvre in August 1572," *L'Esprit Créateur* 54, no. 2 (Summer 2014): 33–44.

14. For the Hôtel de Than, see Bernard Beck, "Les monuments civils de la Renaissance caennaise," in *L'architecture de la Renaissance en Normandie: Actes du colloque de Cerisy-la-Salle*, ed. Bernard Beck (Caen: Presses universitaires de Caen, 2003), 142; and Philippe Lenglart, *Caen—Architecture et Histoire* (Condé-sur-Noireau: Corlet, 2008), 209, 216–18, 221. For the Hôtel de Nollent, see Beck, "Les monuments civils," 142–43; Lenglart, *Caen*, 218–19.

15. On the importance of the *Trionfi* outside of Italy, see inter alia Robert Coogan, "Petrarch's *Trionfi* and the English Renaissance," *Studies in Philology* 37, no. 3 (1970): 306–27. Coogan reminds us that Ascham in *The Scholemaster* (1563–1568) laments that

Italianate Englishmen have "more reverence for the *Triumphs* of Petrarch than for the Genesis of Moses" (306).

16. Patricia L. Corcoran, Charles J. Moore, and Kelly Jazvac, "An Anthropogenic Marker Horizon in the Future Rock Record," *GSA Today* 24, no. 6 (June 2014): 5.

17. The Great Pacific Garbage Patch is of course, for the most part, invisible to the naked eye because it is primarily constituted of plastic microgarbage, yet as a "projection of guilt" as a "trash archipelago" it remains a compelling rallying image, both practically and theoretically. For first thoughts on this, see Daniel Engber, "There Is No Island of Trash in the Pacific," *Slate*, September 12, 2016. http://www.slate.com /articles/health_and_science/the_next_20/2016/09/the_great_pacific_garbage_patch _was_the_myth_we_needed_to_save_our_oceans.html. The state of Kamilo Beach is a concentrated version of this archipelago.

18. Chris Maser, *Interactions of Land, Oceans, and Humans: A Global Perspective* (London: CRC, 2015), 147.

19. Bruno Latour, *Face à Gaïa* (Paris: La Découverte, 2015), 159: "[impossible de] départager l'homme de la nature."

20. Jules Michelet, *Introduction à l'histoire universelle* (Paris: Hachette, 1831), 5. Translation mine. French: "Avec le monde a commencé une guerre qui doit finir avec le monde, et pas avant; celle de l'homme contre la nature [et] de l'esprit contre la matière."

21. Michelet, *Introduction à l'histoire universelle*, 7. French: "Les Alpes n'ont pas grandi, et nous avons frayé le Simplon."

22. There would be much to say about the role played by nineteenth-century scholarship in establishing the Renaissance as a historical moment in which Man is separated from Nature, a wholly mistaken understanding of what early modern thinkers and writers actually did. For now, let us note that Burckhardt, for example, stated that the beginning of modernity coincided with the story of "man's separation from nature," made possible by the Renaissance "awakening of the mind." Christophe Bonneuil and Jean-Baptiste Fressz, *L'événement Anthropocène: La Terre, l'histoire et nous* (Paris: Le Seuil, 2013), 45–46. A full study on the Nature/Culture divide in nineteenth-century readings of the early modern would be most timely and invaluable.

23. The major references for the art and architecture of sixteenth-century France mentioned earlier in this chapter's notes (Lemerle and Pauwels, *L'architecture à la Renaissance*; Blunt, *Art and Architecture in France, 1500–1700*; and Zerner, *L'art de la Renaissance en France*) all assume a clear distinction between "Man" and "Nature." Whether these works aim to be explicitly in debt to Burckhardt and/or Michelet, they arguably still demonstrate this same sense of a clear separation.

24. On this building, see Beck, "Les monuments civils," 145–150; Lenglart, *Caen*, 122–134; as well as Andreas Förderer, "L'hôtel d'Escoville à Caen: essai de restitution de l'état original," in Beck et al., eds., *L'architecture de la Renaissance en Normandie*, 163–74.

25. See Jean-Luc Nancy, *The Ground of the Image*, trans. Jeff Fort (New York: Fordham University Press, 2005).

26. Dugué et al., *La pierre de Caen*, 41. For a more developed discussion of these statues from a similar perspective, see Phillip John Usher, "The Night before Geology:

Fossil Stories from Early Modern France," in *Renaissance Storytelling*, ed. Emily Thompson (Newark: University of Delaware Press, forthcoming).

27. The original edition is Charles de Bourgueville, *Les recherches et antiquitez de la province de Neustrie, à présent duché de Normandie, comme des villes remarquables d'icelle, mais plus spécialement de la ville et université de Caen, par Charles de Bourgueville, . . .; Les Recherches et antiquitez de la ville et université de Caen et lieux circonvoisins des plus remarquables, par Charles de Bourgueville, sieur du lieu de Bras et de Brucourt* (Caen: impr. de J. Le Fèvre, 1588), BnF Rés. LK2-1180 (A). I will be quoting from the following nineteenth-century edition: *Les recherches et antiquitez de la province de Neustrie* (Caen: impr. de T. Chalopin, 1833). Bourgueville was also the author of several other works, most memorably the *Athéomachie et discours de l'immortalité de l'âme et résurrection des corps* (Paris: Martin Le Jeune, 1564). He was also—and this detail will have its importance in due course—a translator: Dares Phrygius, *L'histoire véritable de la guerre des Grecs et des Troyens . . . escrite premièrement en grec par Darès de Phrygie, depuis traduite en latin par Cornille Nepveu et faite françoise par Charles de Bourgueville* (Caen: B. Macé, 1570). All translations mine.

28. See Marie-Madeleine Fontaine, "L'intérêt pour l'architecture chez les écrivains normands: Charles de Bourgueville, Guy Le Fèvre de la Boderie, Jean Doublet et quelques autres," in Bernard Beck et al., eds., *L'architecture de la Renaissance en Normandie*, 59–81.

29. Bourgueville, *Les recherches et antiquitez*, 1:6.

30. Bourgueville, *Les recherches et antiquitez*, 1:7.

31. Bourgueville, *Les recherches et antiquitez*, 1:35.

32. Bourgueville, *Les recherches et antiquitez*, 2:28–29.

33. Bourgueville, *Les recherches et antiquitez*, 2:29.

34. Bourgueville, *Les recherches et antiquitez*, 2:250.

35. Bourgueville, *Les recherches et antiquitez*, 2:176.

36. Bourgueville, *Les recherches et antiquitez*, 2:266.

37. Bourgueville, *Les recherches et antiquitez*, 2:266.

38. Bourgueville, *Les recherches et antiquitez*, 2:171.

39. Among the important works of early modern epigraphy, see, for example, Giacomo Mazochi's *Epigrammata antiquae Urbis* (1521), which collected about three thousand inscriptions from Rome and the surrounding area, as well as Peter Apian's *Inscriptiones sacrosanctae vetustatis* (1534). Two of the period's culminating works are Jan Gruter's *Inscriptiones antiquae totius orbis Romani* (1602–1603) and Jean-Jacques Boissard's *Antiquitates Romanae* (1597–1602). Much has been written about early modern epigraphy. As a starting point, see William Stenhouse, *Reading Inscriptions and Writing Ancient History: Historical Scholarship in the Late Renaissance* (London: Institute of Classical Studies, 2005).

40. Bourgueville, *Les recherches et antiquitez*, 1:5.

41. Bourgueville, *Les recherches et antiquitez*, 1:7.

42. For a list of similar contemporary anagrams, in which Bourgueville's is indeed included, see Prosper Blanchemain, "Catalogue des anagrames, devises & pseudonymes

des poètes du XVIᵉ siècle," in *Miscellanées bibliographiques*, ed. Edouard Rouveyre and Octave Uzanne (Paris: Edouard Rouveyre, 1878), 161–67.

43. Bourgueville, *Les recherches et antiquitez*, 1:3. His image is on the same page.

7. SALINE INTIMACIES

1. The opposition between mined salt (natural) and salt produced by evaporation (artificial) is widespread in the early modern period. See inter alia Georgius Agricola, *De natura fossilium (Textbook of Mineralogy)*, ed. and trans. Mark Choice Bandy and Jean A. Bandy (New York: Geological Society of America, 1955), 36. The ultimately exterranean nature of all salt would not be understood until 1715, when the British astronomer Sir Edmond Halley "suggested that salt and other minerals were carried out to sea by rivers," as noted by John Zumerchik and Steen L. Danver, eds., *Seas and Waterways of the World—An Encyclopedia of History, Uses, and Issues* (Santa Barbara, CA: ABC Clio, 2010), 1:419.

2. The term is from Jeffrey Jerome Cohen, *Stone: An Ecology of the Inhuman* (Minneapolis: University of Minnesota Press, 2015), 21, already quoted at the beginning of Chapter 6.

3. "How Stuff Works: Salt," https://www.youtube.com/watch?v=gI5qV-kvLeg.

4. Bernard Palissy, *Discours admirables de la nature des eaux et fontaines, tant naturelles qu'artificielles, des métaux, des sels et salines, des pierres, des terres, du feu et des émaux* (Paris: Martin le jeune, 1580), 168–69.

5. Pamela H. Smith has studied Bernard Palissy as part of her ongoing investigations into the relationship between artisanal and scientific practices in the early modern period, in "In the Workshop of History: Making, Writing and Meaning," *West 86th* 19, no. 1 (Spring–Summer 2012): 4–31.

6. Umberto Eco, *The Infinity of Lists: From Homer to Joyce*, trans. Alastair McEwan (New York: Rizzoli, 2009), 49.

7. Palissy, *Discours admirables*, 194.

8. Ronsard's "employment of this verse form was on so much vaster a scale [compared to Lemaire, Marot, and Peletier] that he might properly claim credit as an innovator." Isidore Silver, *Ronsard and the Hellenic Renaissance in France* (Geneva: Droz, 1987), 1:448.

9. André Mage de Fiefmelin, *Le saulnier ou de la façon des Marois salans*, ed. Julien Goeury and Nicole Pellegrin (La Rochelle: Rumeur des Ages, 2005), 59.

10. Quentin Meillassoux defines correlation as follows: "Par 'corrélation,' nous entendons l'idée suivant laquelle nous n'avons accès qu'à la corrélation de la pensée et de l'être, et jamais à l'un de ces termes pris isolément" ("By 'correlation' we mean the idea according to which we only ever have access to the correlation between thinking and being and never to either term considered apart from the other"); and correlationism in this way: "Le corrélationisme consiste à disqualifier toute prétention à considérer les sphères de la subjectivité et de l'objectivité indépendamment l'une de l'autre" ("Correlationism consists in disqualifying the claim that it is possible to consider the realms of subjectivity and objectivity independently of one another"). Quentin Meillassoux,

Après la finitude (Paris: Seuil, 2012), 18–19; *After Finitude*, trans. Ray Brassier (London: Continuum, 2010), 5.

11. Quoted from *La vie de Saint-Louis en vers François par personnages*, ed. Francisque-Michel (Westminster: Roxburghe Club, 1871), 170.

12. I quote here Lazare Sainéan, *La langue de Rabelais* (Geneva: Slatkine Reprints, 1976), 1:455.

13. I paraphrase here Élyse Dupras, *Diables et saints* (Geneva: Droz, 2006), 204: "Les ivrognes parisiens . . . subissent les mauvais tours du diable Pantagruel."

14. This is the summary provided by Michael A. Screech, *Rabelais*, trans. Marie-Anne de Kisch (Paris: Gallimard, 1992), 55.

15. Simon Gréban, *Le mystère des actes des apôtres*, electronic ed., CNRS-Lamop (UMR 8589), http://eserve.org.uk/anr/. See also Raymond Lebègue, *Le mystère des actes des apôtres, contribution à l'étude de l'humanisme et du protestantisme français au XVIᵉ siècle* (Paris: Champion, 1929), 242–43.

16. Rabelais, after Annius Viterbo, is generally credited with creating the first likeable giants. For a full history, see Walter Stephens, *Giants in Those Days* (Lincoln: University of Nebraska Press, 1989).

17. Quotes from Rabelais are from François Rabelais, *Pantagruel*, ed. Guy Demerson (Paris: Seuil, 1996), chaps. 28 and 29, 282–305. English translations are adapted from François Rabelais, *Gargantua and Pantagruel*, trans. M. A. Screech (London: Penguin, 2006), 133–45.

18. See Friedrich A. Flückiger and Daniel Hanbury, *Pharmacographia: A History of the Principal Drugs of Vegetable Origin* (London: Macmillan and Co., 1874), 502–4.

19. For a review of the role of salt in Rabelais's texts, see Mireille Huchon, "Le sel rabelaisien," in *Du sel. Actes de la journée d'études "Le sel dans la literature française"* (Biarritz: Atlantica, 2005), 91–113; and Romain Ménini, *Rabelais altérateur. "Græciser en françois"* (Paris: Classiques Garnier, 2014), 882–87.

20. Mikhail Bakhtin, *Rabelais and His World*, trans. Hélène Iswolsky (Bloomington: Indiana University Press, 1984), 340.

21. See the first footnote to the present chapter.

22. Blaise de Vigenère, *Traicté du feu et du sel* (Paris: chez la veufve Abel L'Angelier, 1618), 235.

23. Quoted from "Sel," in Randle Cotgrave, *A Dictionarie of the French and English Tongues* (London: Adam Islip, 1611).

24. For other readings of this *saliera*, see Phillip John Usher, *Epic Arts in Renaissance France* (Oxford: Oxford University Press, 2014), 1–2 and notes.

25. Plutarch, *Symposiacs*, in *Plutarch's Morals*, ed. and trans. William W. Goodwin (Boston: Little, Brown, 1870), 305.

26. Vigenère, *Traicté du feu et du sel*, 240. English translation from Plutarch, *Symposiacs*, 305.

27. The poem was first published in 1601 as part of Fiefmelin's works: *Les œuvres du Sieur de Fiefmelin, divisées en deux parties, conteniies en la page suivante* (Poitiers: Jean de Marnef, 1601). I will quote from the following modern edition: Fiefmelin, *Le saulnier*,

ou de la façon des Marois salans et du sel marin des isles de Sainctonge, ed. Julien Goeury and Nicole Pellegrin (La Rochelle: Rumeur des Ages, 2005). I should like to thank Julien Goeury for introducing me to this poem via his paper given at a conference titled "Iles et insulaires," at the Sorbonne, March 17–18, 2016, and for giving me a copy of his edition. For a full introduction to the poem, see Julien Goeury, "Le sel de la muse," in Fiefmelin, *Le saulnier*, ed. Goeury and Pellegrin, 9–51.

28. Fiefmelin, *Le saulnier*, 55, v. 4–6.

29. Fiefmelin, *Le saulnier*, 57, v. 57 and v. 62.

30. Fiefmelin, *Le saulnier*, 80, v. 703–06.

31. Fiefmelin, *Le saulnier*, 80, v. 707–09.

32. Quoted by G. Dunoyer de Segonzac, *Les chemins du sel* (Paris: Gallimard, 1991), 80.

33. Fiefmelin, *Le saulnier*, 58, v. 67–72.

34. Fiefmelin, *Le saulnier*, 76, v. 592.

35. For ease of reading, I reproduce here Julien Goery's explanations of specific French terms. "*Bris*: terre composite servant à constituer une digue ou un remblai. On appelle *terre de bris* la terre glaise, la vase marine imperméable qui sert à l'établissement des marais; *séeller*: rendre étanche; *tenant*: collant; *platin*: banc de sable, que les basses mers découvrent; *chenal*: canal d'adduction, parfois assez large pour être navigable, qui conduit l'eau de mer, parfois sur plusieurs kilomètres, jusque dans le Jas; *ruisseau*: sorte d'affluent du Chenal qui sert à la fois à conduire l'eau salée vers les Champs et à déverser à la mer les eaux douces qui s'y accumulent; *provigner*: produire en abondance." From Fiefmelin, *Le saulnier*, 107–17.

36. Fiefmelin, *Le saulnier*, 64, v. 235–45.

37. Cotgrave (*Dictionarie*, "Cruche") translates this expression as: "Such stuffe, such work; such wit, such words; such affection, such actions; such discretion, such directions."

38. Olivier de Serres, *Théâtre d'agriculture et mesnage des champs* (Paris: chez Abr. Saugrain, 1603), sig. Av^r.

39. de Serres, *Théâtre d'agriculture et mesnage des champs*, 4.

40. Fiefmelin, *Le saulnier*, 66, v. 283–84.

41. Fiefmelin, *Le saulnier*, 66, v. 287–88.

42. On this point, see Phillip John Usher and Isabelle Fernbach, "Introduction," in *Virgilian Identities in the French Renaissance*, ed. Phillip John Usher and Isabelle Fernbach (Woodbridge: D. S. Brewer, 2012): 1–18; and Annabel Patterson, "Pastoral vs. Georgic: The Politics of Virgilian Quotations," in *Renaissance Genres: Essays on Theory, History, and Interpretation*, ed. Barbara Kiefer (Cambridge, MA: Harvard University Press, 1986): 241–67.

43. Fiefmelin, *Le saulnier*, 63, v. 194–22.

44. I quote the poem ("Ad Janum Terinum de salisfodinis Sarmatiae, quas per funem immissus lustraverat") from the following modern edition: Pierre Laurens, ed., *Musae reduces. Anthologie de la poésie latine dans l'Europe de la Renaissance* (Leiden: Brill, 1975), 1:320–23. For the original edition, see Conrad Celtis, *Quatuor libri amorum secundum quatuor latera Germanie* (Nürnberg: Sodalitas Celtica, 1502), f. xiii^v–xiiii^v. Translations mine. References are to line numbers. Here, 13–14.

45. Jean de Marconville, *De la dignité et utilité du sel* (Paris: Vve J. Dallier et N. Roffet, 1574), 85. Marconville opposes rock salt to salt he calls "artificial, made from the water and from the sea's foam" ("artificiel, faict de l'eau et escume de la mer"). For a general introduction to the history of the salt mines of Wieliczka, see Janusz Podlecki and Stanisław Anioł, *Wieliczka, Historic Salt Mine* (Cracow: Karpaty, 2007).

46. Johann Gottfried, *Salisfodinae Cracovienses Regis Poloniarum Augusti III. pii magnanimi pacifici P. P. admirabili Providentia in tractu Vielicensi restauratae* (s.l.: s.n., 1760), Bibliothèque de l'INHA, Pl Est 63. I should like to thank Benoît Bolduc for this reference.

47. Based on Lewis W. Spitz, *Conrad Celtis: The German Arch-Humanist* (Cambridge, MA: Harvard University Press, 1957), 11–18.

48. Celtis, "Ad Janum Terinum," lines 7–8.

49. Claudia Wiener, "Die Aeneas-Rolle des elegischen Helden. Epische Inszenierung und dichterisches Selbstverständnis in Celtis' *Amores*," in *Vestigia Vergiliana: Vergil-Rezeption in der Neuzeit*, ed. Thorsten Burkard, Markus Schauer, and Claudia Wiener (Berlin: De Gruyter, 2010), 73–105, esp. "Der Gang in die Unterwelt" (77–84).

50. Celtis, *Quatuor libri*, line 25.

51. Celtis, *Quatuor libri*, line 36.

52. John Milton, *Paradise Lost*, ed. Gordon Teskey (New York: Norton, 2005), 1:63.

53. One might take this reading further via Victor Stoichita's *Brève histoire de l'ombre* (Geneva: Droz, 2000), to differentiate more clearly Celtis's looking into total blackness and seeing black stars (which relates to a hyperawareness of self) from the Otherness that Stoichita associated with shade, shadows, and partial light, opposing Narcissus looking at his reflection (*le stage du miroir*) and the reign of shadows (*le stade de l'ombre*).

54. Celtis, "Ad Janum Terinum," line 41.

55. Celtis, "Ad Janum Terinum," line 48.

56. Claudia Wiener, "Die Aeneas-Rolle des elegischen Helden," 81.

57. Celtis, "Ad Janum Terinum," lines 5–6.

58. François de Belleforest, *La cosmographie universelle* (Paris: chez Michel Sonnius, 1575), bk. 4, col. 1804.

59. Celtis, "Ad Janum Terinum," line 43.

60. Agricola, *De natura fossilium*, 36–42.

61. Timothy Morton, *Dark Ecology: For a Logic of Future Coexistence* (New York: Columbia University Press, 2016), 6.

62. Kerry K. Karukstis and Gerald R. Van Hecke, *Chemistry Connections* (San Diego: Elsevier Science, 2003), 29–30.

EXPLICIT

1. David Wallace-Wells, "The Uninhabitable Earth," *New York Magazine*, July 9, 2017. http://nymag.com/daily/intelligencer/2017/07/climate-change-earth-too-hot-for -humans.html.

2. Following the order of "The Uninhabitable Earth," we thus read Wallace-Wells describing, first, how scientists have yet to factor into future predictions what the

impact will be of the "dieback of forests and other flora (which *extract* carbon from the atmosphere)"; second, how the geochemist Wally Broecker thinks that "no amount of carbon emissions reduction" can have a meaningful impact on global warming and how he believes rather in carbon capture, i.e., "untested technology [designed] to *extract* carbon dioxide from the atmosphere"; and, third, how the climate scientist James Hansen no longer believes in the idea of a carbon tax, preferring now to think (like Broecker) about *"extracting* carbon from the atmosphere."

3. Oliver Milman and Sam Morris, "Trump Is Deleting Climate Change, One Site at a Time," *Guardian*, May 14, 2017; Michael D. Shear, "Trump Will Withdraw US from Paris Climate Agreement," *New York Times*, June 1, 2017.

4. Stacy Alaimo, *Exposed: Environmental Politics and Pleasures in Posthuman Times* (Minneapolis: University of Minnesota Press, 2016), 95–98.

5. "An America First Energy Plan," http://donaldjtrump.com/policies/energy. As of June 30, 2017, this page no longer exists. A similar (but not identical) text can be found at "An American First Energy Plan," https://www.whitehouse.gov/america-first-energy.

6. Michael Klare, "Donald Trump Wants to Drown the World in Oil," *Mother Jones*, December 17, 2016, http://www.motherjones.com/politics/2016/12/oil-donald-trump -climate-change-epa-rex-tillerson/.

7. On leaving The Kitchen one evening in 2015 after seeing the final New York performance of *Gaïa Global Circus*, I walked along Nineteenth Street wondering who/ what/where Gaïa was: the balloon I had gone home with, the huge canopy under which I had been seated, the stage, the human actors, the nonhuman objects on stage, or all of the above. In the maelstrom of bodies, colors, and sounds, Gaïa became both impossible to see—yet omnipresent in her soliciting of responses and in the appeals made to her. The first three chapters explored Terra in a similar manner.

8. "Il nous faudroit des topographes qui nous fissent narration particuliere des endroits où ils ont esté" ("We ought to have topographers who would give us an exact account of the places where they have been"). Michel de Montaigne, *Essais*, ed. Pierre Villey and V. L. Saulnier (Paris: PUF, 2004), I, 31, 205; *Essays*, trans. Donald Frame (New York: Everyman's Library, 2003), 184.

9. Bruno Latour, "Anti-Zoom," in *Scale in Literature and Culture*, ed. M. Tavel Clarke and D. Wittenberg (New York: Palgrave, 2017), 101. The question of scale organized my *L'aède et le géographe: poésie et espace du monde à l'époque prémoderne* (Paris: Classiques Garnier, 2018). See especially the conclusion (341–47) to that book, which leads in directions different from the present discussion.

10. Timothy Morton, *Dark Ecology: For a Logic of Future Coexistence* (New York: Columbia University Press, 2016), 12.

11. Bruno Latour, *Où atterrir?* (Paris: La Découverte, 2017), 74.

12. Eugene Thacker, *In the Dust of This Planet* (New York: Zero, 2011).

13. Ashley Alman, "The Opposite of Addiction Is Not Sobriety. The Opposite of Addiction Is Connection," *Huffington Post*, July 9, 2015, http://www.huffingtonpost.com /2015/07/09/drug-addiction_n_7765472.html. I think here, of course, of Bill McKibben, *Eaarth: Making a Life on a New Planet* (New York: Time Books, 2010).

INDEX

Eglise Saint-Pierre de Caen, 117
emission, 1
Erasmus, Desiderius, 7, 81
"Erde an Trump: Fuck you!," 152
Erzgebirge. *See* Ore Mountains
Essais (Montaigne), 9, 16, 61–65, 66, 70–75
Euripides, 85
evaporation ponds, *144, 145*
evil, 32, 38, 109, 111
ex Terra, 19, 27, 32, 155, 160n13; *a Terra* to, 40;
    agriculture, 33; ethical constraints on
    extracting, 35; extraction of metal, 43;
    mining, 39; price of extracting, 83
Excidio Troiae (Phrygius), 129
exterranean activity, 3; Agricola and, 80;
    close up perception of, 79; Fortuna
    response, 65; globe and, 61; Jupiter and,
    47; Klein and, 45; labor, 69; materiality
    of, 32; as matricidal, 27, 29; of moles, 78;
    moments in process of, 155; Petrarch
    and, 38; Terra and, 61, 74; uninhabitabil-
    ity of Earth, 152; violence of, 26, 44
extinction rates, 5
extraction, 1, 12, 86; agency and, 61;
    antagonistic nature of, 89; artisanal,
    54; Caen, stone extraction in, 120; *De re
    metallica* and, 80, 87; deforestation and,
    92; destructiveness of, 98; exporting
    of, 74; of gold, 65; human agents of, 85;
    large-scale, 66; of metal *ex Terra*, 43;
    need to rephenomenalize, 3; Pliny on,
    27; of salt, 142; sites of, 48
*Eyn wohlgeordnet und nützlich büchlein wie
    man bergwerk suchen und nden soll* (von
    Calw), 44–45

Face à Gaïa (Latour, B.), 39
Fachs, Modestin, 11
fertility, 92, 107
FeS$_2$. *See* pyrite
Feuerbach, Ludwig, 8
Fever Room (Weerasethakul), 78
Fiefmelin, André Mage de, 114, 136, 137, 143,
    147, 149
Fight with Cudgels (Goya), *93, 94*

Florilegium (Stobaeus), 49
follets, 108
Fortuna, 40
Fortune, 36; Jupiter and, 38; Niavis and, 37;
    Wheel of, *37*
fossil fuels, 5
fossil futures, 115
France, 122, 142; art and architecture of
    sixteenth-century, 192n23; Atlantic
    coast of, 143; drought in, 141; galena
    and, 108; geography and geology of,
    109, 191n9; humanism and, 8; "Hymne
    de France," 48; mining in, 106, 108;
    mining industry in, 106, 107; once under
    water, 117; Ramus in, 82, 85; silver in,
    109; *Terre* and, 48. *See also* Caen
Francis (Pope), 179n5
François Ier, 120, 121, 142, 154
Fraysse, Cécile, 78
Freeing of Andromeda, 125
French Renaissance art, 50
fresh water, 5
Fresnaye, Jean Vauquelin de la, 117
Freud, Sigmund, 83
Fuchs, Leonhart, 88

Gaia, 17; of Homeric and Orphic hymns,
    45; metaphor of, 40; motherhood of, 26;
    Terre and, 54
Gaïa Global Circus (play), 21–22, *22*
galena, 108
Garcia, Tristan, 14
Garrault, François, 19, 79, 99, 106; Agricola
    and, 107; on *koباloi*, 109
gemstones, 145
genocide, 65
Geographia (Ptolemy), 57
geography, 58
geological time, 115, 190n1
geology, 115
geomedia, 120, 129, 132, 156
Gesner, Conrad, 49
Ghini, Luca, 56
Giustiano, Vito R., 8
glarea (gravel), 102

Morton, Timothy, 14, 155, 174n66; on
  ecognostic jigsaws, 150; on Romantic
  Nature, 26
motherhood, 26, 35; biological maternity,
  29; Terra Mater as mother, 28
Mundus subterraneus (Kircher), 99
Münster, Sebastian, 57
Mystery Play about the Acts of the Apostles
  (Gréban), 139

NaCl. *See* sodium chloride
Nancy, Jean-Luc, 125
Natura, 29; Agricola and, 87; Park on, 32;
  Terra and, 30, 32, 35, 43
Natural Contract (Serres), 29, 93
natural gas, 141
Natural History (Pliny), 27
naturalia signa, 96
Nature, 93; Agricola on, 86–87, 92; *daemones*
  and, 110; Garrault on, 107; lack in, 43
La nature et les prodiges (Céard), 1
Naumachius, 49, 85
Navarre, Jeanne de, 48
Neptune, 142, 146, 150
Nevers, 108
New World, 12, 55, 70; anthroturbation of,
  66; "discovery" of, 57, 59; giving voice
  to humans of, 73; gold from, 49; mas-
  sacre of populations in, 61–62; mining
  in, 67; Old World and, 65; terraforma-
  tion in, 71
New York Magazine, 152
Niavis, Paulus, 22–23, 24, 25, 30, 36, 39, 86;
  Fortune and, 37; mining and, 70; Ovid
  and, 35; Petrarch and, 38; Physis and, 40;
  Pliny and, 27
The Night of the Moles (Quesne), 18, 78, 78,
  153, 154
Nollent, Gérard de, 121
Normandy, 116, 117, 120
Nymphéas (Monet), 2
Nymphéas transplant (14–18) (Huyghe), 2

object-oriented ontology (OOO), 14
Old World, 65

Oléron, 143; evaporation ponds, 144; salt
  workers of, 146
"On Coaches" (Montaigne), 61
OOO. *See* object-oriented ontology
Orbis hypothesis, 12, 13, 166n66, 167n69
ore, 109
Ore Mountains, 24, 29, 34, 94; Annaberg
  mine in, 102
Orphic Hymns, 26, 45
osmosis, 134
Où atterrir? (Latour, B.), 155
Ovid, 33, 34, 35, 42, 45, 70, 85; Agricola and,
  86; early ages of, 71

Pagan gods, 47; Ceres, 32, 33, 34, 142, 146;
  Mercury, 24, 25, 47; Neptune, 142, 146,
  150; Proteus, 140; Triton, 140. *See also*
  Fortune; Jupiter
Palissy, Bernard, 134, 135, 137, 149
Pantagruel (fictional character), 138, 139;
  Loup Garou and, 141; salt and, 140
Pantagruel (Rabelais), 138, 139
Paracelsus, 19, 79, 110, 111, 112
Park, Katherine, 32
Parte primera de la chronica del Perv
  (de León), 66, 67, 67
Peru, 144
Petrarch, 7, 38, 121, 122, 122–23
petroleum, 141
Phocylides, 85
Phrygius, Pseudo-Dares, 129
Physis, 26, 39, 40
De planctu naturae (de Lille), 30, 32
Plastiglomerate, 124
Pliny, 27
Plutarch, 142
pneuma, 102
pollution, 33
posthumanism, 5, 7, 8, 151
Posthumus, Stephanie, 14
Potosí, 66, 67, 68, 74
Probier Büchlein (Fachs), 11
Probierbüch (Zimmermann), 11
Probierbüchlin (Schreittmann), 11
prospectors, 94, 95, 96

Protestants, 106, 129, 132
Ptolemy, 57
Puerto Rico, 74, 181n23
pulmonary problems, 104
Purdy, Jedediah, 151
pyrite (FeS$_2$), 53

quarries, 127
Queen Elizabeth I (Hilliard), *31*
Quesne, Philippe, 18, 77–78

Rabelais, François, 19, 114, 139
Rape of Europa, 125
Reason, 38
Recherches et antiquitez de la province de Neustrie (de Bourgueville), 126, 128, 129, 131
Recherches et antiquitez de la ville et université de Caen et lieux circonvoisins (de Bourgueville), 126, 128, 129, 131
removal, action of, 3
Reset Modernity! (Latour, B.), 2
Robertson, Robert, 55
Romantic Nature, 26
Ronsard, Pierre de, 41, 42–43, 45, 48, 49, 50–51, 70, 136; Jupiter and, 46; *La salade*, 147

Saint Bartholomew's Day Massacre, 106
St. Joachimsthal, 81
La salade (Ronsard), 147
saliera (Cellini), 114, 142, *143*
salt, 136, 155; aspects of, 135; extraction of, 142; functions of, 134; limestone and, 133; mines, 147–49; Oléron, salt workers of, 146; origins of, 141; Pantagruel and, 140; production of, 143, 144, 194n1; rock, 144; salt water, 133; types of, 133; weaponized, 141
Salzbergwerk Dürrnberg, 149
Le saulnier, ou de la façon des marois salans et du sel marin des isles de Sainctonge (de Fiefmelin), 136, 143, 147
sedimentary rocks, 133
Self-Tormentor (Terence), 9

Sententiae ex thesauris Graecorum delectae (Gesner), 49
Serres, Michel, 1, 14, 29, 93, 98
Serres, Olivier de, 146
shaft houses, 89
Sherman, William, 96
silicosis, 111
silver, 60, 68, 108, 109
smelting, 109
Smith, Pamela, 88
Socrates, 84
sodium chloride (NaCl), 133, 134, 139
Stiefvater, Maggie, 8
Stobaeus, Joannes, 49, 85
Stoermer, Eugene, 5
stone, 115, 129, 131, 132
Stone: An Ecology of the Inhuman (Cohen), 15–16, 114, 115
subterranean bodies, 100–1; subterranean animals, 103, 188n10
Symposiacs (Plutarch), 142
syphilis, 56

taxes, 144
teichoscopia, 127
Telesio, Bernardino, 46
Tellus, 160n11. *See also* Terra
Terence, 9
Terinus, Janus, 147
terra, 97; Terra and, 77, 154
Terra, 16, 18, 24, 26, 55, 97, 155, 156; bodiliness of, 27, 38; *Cosmographia* and, 60, 64; divisions of, 74; exterranean activity and, 61; figures of, 57; geography and, 58; as globe, 59, *59*; human relationship to, 75; "Hymn to Gold" and, 42; large-scale harmonious relationship with, 34; Merchant on, 174n66; mining of, 29; multiple "things" of, 21; Natura and, 30, 32, 35, 43; terra and, 77, 154; Terra on the Ara Pacis, Rome, 28; wounding of, 25
a Terra, 40
Terra Mater, 26, 107; as mother, 28
terraformation, 33, 34; anthropogenic, 72; modes of, 73; in New World, 71

MEANING SYSTEMS

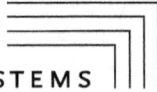

www.ingramcontent.com/pod-product-compliance
Lightning Source LLC
Chambersburg PA
CBHW032135020426
42334CB00016B/1166